T0202180

INTRODUCTORY MATHEMATICS AND STATISTICS THROUGH SPORTS

Introductory Mathematics and Statistics through Sports

Supplementary Activities
and Writing Projects

TRICIA MULDOON BROWN
Georgia Southern University

ERIC B. KAHN
Bloomsburg University

OXFORD
UNIVERSITY PRESS

OXFORD
UNIVERSITY PRESS

Great Clarendon Street, Oxford, OX2 6DP,
United Kingdom

Oxford University Press is a department of the University of Oxford.
It furthers the University's objective of excellence in research, scholarship,
and education by publishing worldwide. Oxford is a registered trade mark of
Oxford University Press in the UK and in certain other countries

Published in the United States of America by Oxford University Press
198 Madison Avenue, New York, NY 10016, United States of America

British Library Cataloguing in Publication Data

Data available

Library of Congress Control Number: 2019933928

ISBN 978–0–19–883567–7

DOI: 10.1093/oso/9780198835677.001.0001

Printed and bound by
CPI Group (UK) Ltd, Croydon, CR0 4YY

Preface

The market is saturated with content-focused textbooks for introductory general education mathematics and statistics classes, but collections of resources to supplement and reinforce the content of these courses are decidedly lacking. The student population that typically takes these courses is composed of non-STEM majors from a wide array of disciplines who often are fulfilling general education requirements. The onus is on faculty to motivate and inspire these students to learn and appreciate the importance of mathematics to their preferred fields of study, but doing so can be a significant challenge.

The authors' two main goals in writing this text were to provide a resource for faculty to tap into students' interest in sports as a tool for learning mathematics and to battle the commonly held position that math is to be feared and disliked. Our approach is simple: use sport as a medium for teaching mathematics through the process of research, oral communication, and writing. By utilizing students' interest in sports as a tool for learning mathematics, we believe that faculty can undermine some of the stigma surrounding mathematics. Students in the fields of education, the liberal arts, and business often have honed their research, communication, and writing skills in their major classes, and by applying their strengths to mathematics content, a more positive learning environment can be created.

Perhaps the most significant benefits for students who learn mathematics with this text is the heavy use of inquiry experiences. While inquiry-based learning (IBL) has been historically associated with upper-level mathematics, the mathematical community has grown to accept that IBL encompasses a broad range of active learning strategies at multiple levels, and that the diverse benefits from learning in an IBL setting are well documented. One of the most common concerns faculty have about switching to an IBL teaching style is a fear of failing to cover necessary content for courses that are prerequisites. While this concern is understandable and discussions about its validity are ongoing, this text offers a solution within the settings of math for liberal arts, introductory statistics, and math content for preservice elementary education classes by allowing faculty to teach material for their courses through any desired means. As the text is supplementary in nature, students in a lecture format classroom will get to experience IBL episodes and the benefits from them while students in an IBL format classroom that incorporates this text will gain from the connections of mathematics to their fields of discipline and topics of everyday interest in their lives. The authors believe it is a strength of this text that faculty do not have to give up their teaching style to utilize these activities and projects—rather, lecturers and IBL faculty can each use the activities and projects to supplement and enhance their classrooms.

In addition, the projects in this text all require the students to effectively communicate their mathematical solutions through writing. The benefits of learning mathematics through writing may only become obvious after questioning why so many departments, some even offering majors seemingly disjoint from mathematics, require students to complete specific mathematics classes. Incorporating writing into the process of learning mathematics allows these students to utilize skills developed in classes from these other disciplines. Further, regardless of how much information one person knows, no matter how elegantly she can solve a problem, without the skills to communicate her knowledge to another human being, her potential to contribute will go unrealized. Perhaps another, potentially less obvious, benefit is that learning to communicate mathematics is also an effective way to learn mathematics; thus, students develop a bond between the skill of abstraction and that of their field of study through communication. The skill of abstracting a problem to more easily work out solutions is not unique to mathematics but certainly is inherent in the study of mathematics.

The activities and projects within this text have been written, classroom tested, and revised over a span of 9 years. In a general education, math for the liberal arts course at Bloomsburg University, Dr. Kahn has consistently used these activities to supplement lecture once or twice per week as one way to bring learning through inquiry into his classroom. In this same class, research and writing assignments have been a more recent addition over the last few years and many have developed into the projects found in this text. Dr. Brown utilizes both projects and activities in her introductory statistics and quantitative reasoning courses, both online and face-to-face at Georgia Southern University (formerly Armstrong State University). In particular, many of these materials were inspired while teaching quantitative reasoning courses linked with a first-year seminar. The IBL-style activities help hold the attention of less disciplined freshman students, and the writing and research required for the projects provide preparation for the higher academic expectations at the college and university level.

We would like to thank all of our students who have worked through the varying versions of the activities and projects that developed into the current forms you see here. Students' insights, pitfalls, and concerns have greatly improved all of these activities from their initial forms; working with students and watching them learn mathematics is part of what makes teaching the subject such a great experience for the authors. We would be happy to receive comments concerning the activities and projects in this text; please direct all comments and inquiries to tmbrown@georgiasouthern.edu and ekahn@bloomu.edu.

Dr. Brown would like to thank her family for their confidence, and especially her husband, Matt, for all of his time and patience. Dr. Kahn would like to express his full gratitude to his wife, Emily, for her unwavering support and encouragement while he was working on this text. Dr. Kahn also would like to thank his daughter Allison, who already is enjoying a positive relationship with mathematics at the age of only 4 and whose love of learning inspires him as a teacher and scholar.

Savannah, GA
Bloomsburg, PA

Tricia Muldoon Brown
Eric B. Kahn

Contents

1 Introduction

In this chapter, we describe the structure of the text, give some general comments on implementation of the materials, and provide some specific examples of how we have utilized the activities and projects in our own classrooms.

1.1 Structure

It is important to remember that this text is designed to be supplemental in nature and is not a replacement for a standard textbook. The material is organized into chapters by mathematical content; within each chapter are five classroom activities and three out-of-the-classroom projects. The activities and projects are all independent of each other and can be utilized throughout a semester in any order. In particular, we do not recommend proceeding straight through the text, as the activities and projects are designed to cover content from courses with various subject matter for students with varying skill sets.

Each chapter starts with a list of mathematical topics covered in the chapter, followed by descriptions of the sports used as a framework for the content. The mathematical list is to be viewed as an aid in determining where the activities and projects most appropriately fit into a course. The sports descriptions are provided to give a minimal introduction to faculty and students unfamiliar with any particular sports.

At the end of the text, you will find two indexes and one glossary. The indexes organize the activities and projects first by mathematical content and then by sport. These allow faculty or the interested student to easily find materials associated with the desired mathematical content or a particular sports model. The glossary is designed particularly for people who have little experience with sports but want to try out some of these activities and projects. It offers a quick history of the origins of each sport and its governing body, in addition to citing websites of official organizations associated with each sport where one can find more information about the specifics of the sport, such as rules, competitions, and historical records.

1.2 Implementation

Although some activities do develop concepts from the ground up, the vast majority assume familiarity with the mathematical content. We envision faculty in a traditional classroom will use an activity to model and reinforce topics first discussed in the immediately previous lecture or, if they are teaching in a flipped classroom, studied the night prior. Having these inquiry experiences embedded early in the process of seeing new material allows students to actively involve themselves in the content and requires them to develop their technical communication skills alongside their mathematical skills rather than after the fact. An active learning experience leads to deeper understanding and longer retention, and by combining the development of communication and content, faculty can encourage a well-rounded student development. Without the ability to express one's technical ideas clearly to others, an individual's mathematical knowledge and the ability to model real-world problems becomes irrelevant.

As the projects all require a significant amount of writing, and many some research, they are intended to mainly be utilized outside the classroom. Within each chapter, the projects are designed to be approximately equal in difficulty and length. The authors have found success in allowing students to read all projects from a chapter and then choose individually which one to complete; by giving this level of control to each individual student, they allow studens to take better ownership of their learning in the course and choose a sport in which they have personal interest. Although there certainly are many ways to integrate projects into a semester, the projects in this text lend themselves to three natural implementations. At times the authors have assigned a chapter's projects immediately following an exam on the same material. This timing provides faculty with flexibility when writing exams, knowing that the material is also going to be formally assessed in a project. If exams covered multiple chapters from the required textbook, the authors have assigned a set of projects immediately following completion of a chapter and as they began the next one. In this scenario students are reviewing the completed material, essentially in preparation for a future exam, while new material is being introduced in the classroom through lectures and activities. In another option, projects are assigned at the beginning of the unit associated with the material. Students can think about how new content and skills could be incorporated into the project as they learn, and assigning a due date on or before an exam allows students to strengthen their skills for that assessment.

Finally, many of the activities and projects contain specific information with regard to teams, dates, and statistics. While we certainly expect the materials to be sometimes implemented exactly as is, when appropriate, we encourage faculty to think of these as templates which can be adjusted for relevancy to student population, geography, and currency. For example, the team data in the activity in Section 4.3.3 could easily be replaced with another team's information more relevant to your area of the country, different statistical measures of pitching success could be used in the project in Section 7.4.2, or the activity in Section 3.3.2 could be updated to contain the current year's MVP candidates. These small changes to help the students connect with the material on a more personal level can make the activity or project more fun and engaging.

1.3 Sample Rubrics and Grading

To conclude, each author has provided personal examples here of how we have incorporated the content from the following chapters in our own courses.

1.3.1 Dr. Brown

Most often I use these materials in a class called Quantitative Skills and Reasoning at the Armstrong Campus of Georgia Southern University. The school is a large, comprehensive, public university, but the campus where I teach is smaller, with about 7,000 students, many of whom are first-generation, nontraditional, or military-affiliated transient students. Much of the time this quantitative skills course is linked with a first-year experience course, the implications being that the size of the math class is smaller and it is entirely composed of freshmen. Generally, I prefer to begin each face-to-face class with a "minilecture." These are about 10 to 15 minutes in length, introduce the day's concepts with some definitions and examples, and are usually accompanied by a slide show of ten or fewer slides. Afterwards, the students get into groups and work through the activity supporting the new material just introduced. With any population of students, but especially with freshmen, I feel that this style of class allows them to remain engaged with the mathematics. I collect the activities every class and correct mistakes, but the grade is solely for participation and only accounts for a small percentage of the overall grade. In these classes I also utilize the projects as out-of-class assignments. I allow the students to work alone or in pairs, and they complete four of their choice throughout the year. These are worth a higher percentage of their course grade and are graded with a rubric such as the one shown below, which assesses the project in Section 2.4.1.

	Inadequate	Needs improvement	Meets expectations	Exceeds expectations
	1	2	3	4
Graph: Correct vertices				
Graph: Correct directed edges				
Graph: Correct weights				
Shortest Hamilton path				
Execution of shortest-path algorithm				
How long?				
Comparisons with traditional schedule				
References				
Spelling, grammar, punctuation				

By using a rubric and allowing pairs to turn in one assignment, the length of time needed to grade these assignment can be made manageable. Note that most projects require a reference list with appropriate formatting and in-text citations, in addition to correct grammar and punctuation, and I expect the students to address all parts of the multipart assignment: these are things many freshman students are not immediately prepared to do, but are necessary skills for college students. With this student population I feel that I grade extremely leniently. I do not expect high-level sentence structure or complex critical thinking. I allow them to use obvious sources without requiring depth and variety and do not expect strong synthesis of the ideas from their sources. Most students pay attention to their feedback, so I am often quite satisfied and sometimes surprised with their improvement from the first to the last project.

1.3.2 Dr. Kahn

I teach at Bloomsburg University, which is a regional, public university of approximately 10,000 students. My general education, math for the liberal arts classes are capped at 40 and usually start out with that many students enrolled. This size can become prohibitive to consistently grading in-class inquiry activities, but I think feedback is especially important to this population of students. To be able to give feedback to everyone, I have the class work on the activities in small groups, and I visit each group, asking for explanations of their work as they progress and clarifying any points of confusion. If a group finishes early, I sit down with them and have each student explain at least one part of the activity and the group's solution. I have found that after students realize I am going to do this, individuals begin to focus on explaining their solutions to all group members, thus improving their communication skills and their peers, content knowledge. I express to students at the beginning of the semester, and periodically remind them throughout the semester, that how they interact with their group members during activities is taken into consideration when assigning points to their participation component of the course grade at the end of the semester.

In this same class, I typically assign three or four projects during the semester, which may be completed individually or with a partner. As these assignments take significantly more effort to complete than the activities, I do grade each project. I have found that using a course rubric makes the grading process easier for me, and students appreciate knowing in advance how they will be assessed. My rubric is intentionally coarse to give me the ability to assign intermediate grades if necessary. I also still give direct comments on the papers students turn in but am freed from trying to discern how many points to give or deduct for each comment; my ability to grade efficiently and, more importantly, consistently, has improved with this system. Below is a sample rubric I use for the projects in my chapter on voting theory; all projects within a given chapter are graded on the same rubric and are out of a total of 35 points.

Letter	Effectively accept the job and summarize the results found in the report.
Score 5	The letter is typed and perfectly clear, and accurately gives an overview of your preference list ballots and voting procedures.
Score 4	A bit too brief, is missing an important item, or contains grammatical errors.
Score 2	Unprofessional, poorly written, confusing, or fails to accurately describe an overview of the ballots and voting procedures.
Score 1	No letter.
Voting process	Effectively complete voting processes for each method required.
Score 20	All votes are done correctly on the correct preference list ballots.
Score 15	The voting processes are completed correctly; any mistakes, if present, are minor in nature and due to the creation of one of the ballots.
Score 10	One voting process has a fundamental error in how the winner was decided.
Score 5	Multiple voting processes are incorrect or there are significant errors in the construction of the preference list ballots.
Score 0	No accurate voting methods are completed.
Analysis	Effectively analyze and explain all voting processes.
Score 10	All analysis and explanations are correct and thorough.
Score 8	The analysis and explanations are correct but could use additional information.
Score 7	Significant room for improvement. Includes multiple errors in analyzing vote results.
Score 3	Little work in analyzing vote results.
Score 0	No analysis.

1.4 In Summary

Thus we have concluded the preliminary information. The following seven chapters contain the classroom resources for implementation in your introductory mathematics classes, followed by the noted indexes and glossary. We hope you find the information, activities, and projects found in this text useful in your classes and inspiring in your lesson planning. Most of all, enjoy and good luck!

2 Graph Theory

This chapter contains five activities and three projects designed to cover introductory graph theory topics, including Hamilton and Euler paths and circuits, shortest-path and weighted tree algorithms, and scheduling by graph coloring. The sports cycling, skiing, and the Olympics provide a framework for the material, but no specialized sports information is needed.

2.1 Mathematics

Graph Theory Terminology

- Subgraphs
- Hamilton paths and circuits
- Euler paths and circuits

Efficient Routes and Shortest Path

- Chinese postman problem
- Dijkstra's algorithm

Spanning Trees and Weighted Search Algorithms

- Breadth-first search
- Kruskal's algorithm
- Prim's algorithm
- Reverse-delete algorithm

Graph Coloring

- Chromatic number
- Applications to scheduling

2.2 Sports

Cycling The most well-known event in professional cycling is the Tour de France, held every year in July. The 21-stage route varies from year to year, but the stages are chosen to a encompass a mix of mountain and flat stages as well as two short team time trials. Distances can be as short as 30 km in the time trials and up to over 200 km on some of the flat stages. The athletes ride through cities and the countryside, even veering out of France into other countries such as Spain or Belgium, but they always conclude by riding into Paris down the Champs-Élysées on the final stage of the Tour. Two activities are related to route planning for the Tour. There are several other events during the yearly professional cycling season, and the final project investigates scheduling concerns for a collection of European races.

Skiing Skiing and snowboarding resorts are popular winter vacation locations, but many resorts are also utilized in the summer for hiking, bicycling, or other outdoor activities. Skiing and hiking maps can be modeled by directed graphs in the case of ski resorts, as the lifts are directed up the mountain and the trails are directed down the mountain, while hiking trails can support an undirected graph, as vacationers an travel in either direction along a trail. Directed graphs and Euler path applications are used in an activity and a project in order to plan vacations to ski resorts. A second activity utilizes weighted search algorithms to efficiently design a new resort.

Olympic Games Hosting an Olympic Games is a massive undertaking, and in recent years the preparedness of the host sites has come under question, with issues including between contaminated water and barely completed stadiums in Brazil, and security concerns and unfinished lodgings in Russia. While both of these events were extremely successful, future hosts will work to be as prepared and have their games as efficient as possible. An activity and a project, respectively, address optimization issues for the 2014 Winter Olympics in Sochi and the 2016 Summer Olympics in Rio de Janeiro. In the activity, students work to schedule television interviews for all the medal-winning US athletes in the winter games of 2014. The writing project asks the students to efficiently prepare for the Olympic torch to travel through the major cities of Brazil.

2.3 Activities

2.3.1 Tour de France City Tour

Content Elementary graph theory vocabulary, Hamilton paths and circuits, Euler paths and circuits, Chinese postman problem.

Activity The Tour de France is a multiday road bicycle race that is held each July in France. Teams of cyclists compete for various honors, including the yellow jersey worn by the overall race leader at the end of each day's stage. As it is one of the most

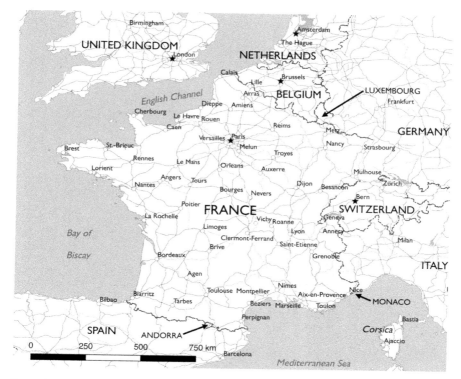

Figure 2.1 Map of major roadways in France

well-known races in cycling, preparations for the 21-stage Tour begin far in advance. This year, organizers are beginning to plan the route for the next race. In particular, the director needs to visit each city and town in order to meet with city planners about the requirements for the race. Use the map found in Figure 2.1[1] to answer the following:

1. Draw a subgraph of roads which would allow the organizers to visit the cities Paris, Rennes, Orleans, Caen, and Nantes. Note that if your road goes through another city, the subgraph must have that city as a vertex.

    (a) Find a route to visit each of the cities on the map. Is a Hamilton circuit possible?

    (b) Label the degree of each vertex. Is an Euler path or circuit possible? If so, find one. If not, explain.

2. Draw a subgraph of roads which would allow the organizers to visit the cities Toulouse, Clermont-Ferrand, Montpellier, Lyon, and Nice.

[1] Mapswire. "Free Maps of France." https://mapswire.com/countries/france/ (accessed: November 29, 2018).

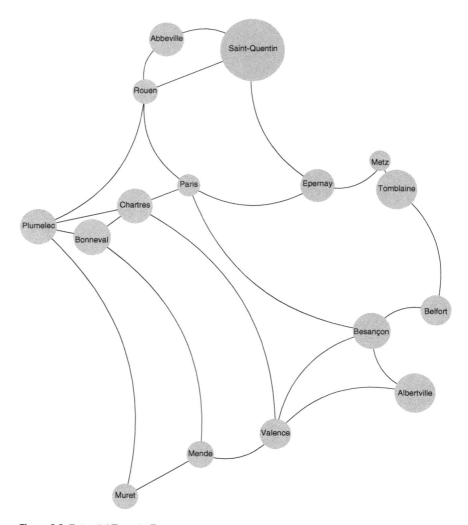

Figure 2.2 Potential Tour de France routes

(a) Find a route to visit each of the cities on the map. Is a Hamilton circuit possible?

(b) Is an Euler path or circuit possible? If so, find one. If not, explain.

3. Find an efficient route that visits each of the cities in Figure 2.2, repeating the fewest number of cities. Did you find a Hamilton path or circuit? Explain using the definitions of a Hamilton path and circuit.

4. Find an efficient route that inspects each of the roads in Figure 2.2, minimizing the number of roads which are repeated. Did you find an Euler path or circuit? Explain using the definitions of an Euler path and circuit.

Implementation Notes Students have a lot of choice for their subgraphs in Questions 1 and 2. Many of them will choose a simple circuit, so if you wish to force them to choose something more complicated, you may need to impose additional criteria. This choice can also be frustrating for some students who are more used to one "right" answer, but a nice feature of graph theory is the multitude of correct solutions.

2.3.2 Tour de France Inspection

Content Route inspection and the Chinese postman problem, Euler paths and circuits, Dijkstra's shortest-path algorithm.

Activity The Tour de France is the premier road bicycle race, and is held over three weeks each July. Cyclists compete as teams for various team and individual honors, and preparations for the Tour begin far in advance. This year, organizers are beginning to plan the route for the next race. They need to inspect the roads which are potential legs for the upcoming race, with the future route being chosen from among the roadways which have been raced in previous Tours. Figure 2.3 shows the remaining roads to be inspected.

1. Assuming any town may be your starting point, find an efficient way to inspect all of these roads and return to your starting point, repeating the fewest number of edges possible. Optimize this route with respect to the distances.
2. What if you are not forced to return to your starting point? Discuss how your route in Question 1 would be different.

After the road inspectors complete their task, two of the inspectors find themselves in Muret and two in Albertville.

3. Use Dijkstra's algorithm to find the shortest way home from Muret to Saint-Quentin.
4. You aren't sure where the remaining inspectors in Albertville live, but you know it is one of the cities on the map in Figure 2.3. Use Dijkstra's algorithm to find the shortest distance from Albertville to every other city on the map.

Implementation Notes You may want to provide extra copies of Figure 2.3 so the students may complete the algorithms multiple times directly on the graph.

2.3.3 Ski Resort Search

Content Spanning trees and breadth-first search, weighted search algorithms—Kruskal, Prim, and reverse-delete.

Activity Skiing and snowboarding are popular recreational activities around the world, and now a new resort is preparing to open. The future layout of the trails is illustrated

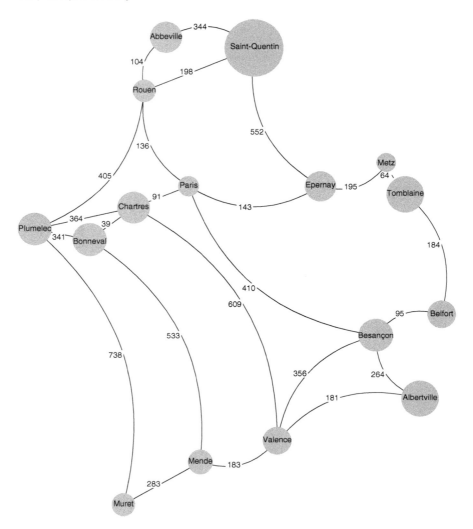

Figure 2.3 Potential Tour de France routes with weights

in Figure 2.4. The project will soon be out of the design phase and into the construction phase. As the whole mountain is currently forested, decisions will need to be made about where to place communication and electrical networks.

1. The first task is to establish communication between each region of the resort. Specifically, the hotel and condo complexes, the Echo, Blizzard, and Pine Tree Lodges, and the Tiptop Tower all need to be in communication, as well as the six intermediate locations indicated on the map. The main receiver at the peak of the mountain will receive the signal and transmit, through a pair of antennas, to each

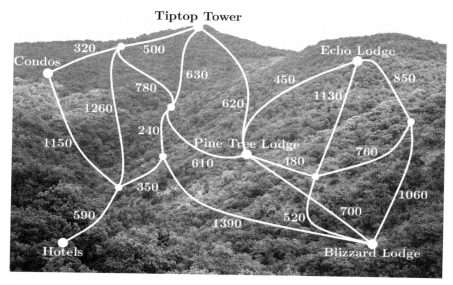

Figure 2.4 Future layout of trials

secondary receiver in a sight line. These receivers, in turn, can transmit through more pairs of antennas to others until a network is set up. To avoid interference from foliage, the signals will follow the trail map shown in Figure 2.4. Use a breadth-first search vertex labeling to connect each area of the resort. How many levels of receivers are needed to reach each region of the resort?

2. The next task is to run electricity to the six resort complexes and the six intermediate locations. The resort owner is very conscious of cost and environmental concerns, so rather than cutting down more trees than necessary, power will be run along existing slopes or lifts. Furthermore, as the costs increase with linear footage, the minimum amount of wires should be run in order to connect all areas of the resort. We need to determine which slopes should be cleared first to prevent a delay in running the electrical supply. We need to experiment with different minimum-cost algorithms to solve this problem.

(a) Use Kruskal's algorithm to determine where the power lines should be placed.

(b) Use Prim's algorithm to place the power lines.

(c) Use the reverse-delete algorithm to lay the power lines.

3. Write a paragraph discussing these three algorithms. Did you get the same tree for each? How are the algorithms the same as or different from each other? Which do you prefer and why?

2.3.4 Euler's Ski Trip

Content Directed graphs, Euler paths.

Munising Ski Trails

➤ Direction of travel

P Parking

Snowmobile trail - Caution!

Mileages and Difficulty
A   2.4   More difficult
B   1.9   Most difficult
C   1.4   More difficult
D   1.0   Easiest
E   8   Easiest
F   1.7   More difficult
G   1.5   Easiest
H   1.0   More difficult
     11.7   miles

Figure 2.5 Map of Munising Ski Trails

Activity Figure 2.5 shows a trail map, created by the National Parks Service[2], of the Munising Ski Trails, located in Pictured Rocks National Lakeshore in Michigan. In this activity, you will plan a route to allow you to see the entire park as efficiently as possible.

1. Convert the trail map into a digraph, with vertices placed where any two trails intersect.

2. Choosing one of the two trailheads, can you find a route following the directions which would allow you to ski every trail in the park? Eulerize this route so it is as efficient at possible.

3. Suppose the directions were removed. Could you find a shorter route to ski over all the trails? Explain.

Implementation Notes This activity is similar to the I.X. Ploor project found later in the chapter, but here the trail map is provided. For a decreased level of difficulty, you may provide the map with the edges or vertices already included.

2.3.5 Olympic Scheduling

Content Complete graphs, graph coloring.

Activity Organizing a Winter Olympics is a massive undertaking. For example, the 2014 games held in Sochi consisted of 7 sports, 15 disciplines, and 98 different events. All of these competitions, including the preliminary rounds, finals, and medal ceremonies, were scheduled in just over two weeks. The media coverage was also hectic. In particular, NBC wanted to schedule interviews for the 28 medalists from the United States. The medalists are found in the table below.

Sport	Discipline	Athlete	Medal
Alpine skiing	Women's slalom	Mikaela Shiffrin	Gold
Alpine skiing	Men's giant slalom	Ted Ligety	Gold
Alpine skiing	Men's super G	Andrew Weibrecht	Silver
Alpine skiing	Women's super combined	Julia Mancuso	Bronze
Alpine skiing	Men's super G	Bode Miller	Bronze
Freestyle skiing	Men's ski halfpipe	David Wise	Gold
Freestyle skiing	Women's ski halfpipe	Maddie Bowman	Gold
Freestyle skiing	Men's slopestyle	Joss Christensen	Gold

[2] US National Park Service. "Munising Ski Trails." National Parks Maps. http://npmaps.com/wp-content/uploads/pictured-rocks-munising-ski-trails-map.jpg (accessed: November 29, 2018).

Sport	Discipline	Athlete	Medal
Freestyle skiing	Men's slopestyle	Gus Kenworthy	Silver
Freestyle skiing	Women's slopestyle	Devin Logan	Silver
Freestyle skiing	Men's slopestyle	Nicholas Goepper	Bronze
Freestyle skiing	Women's moguls	Hannah Kearney	Bronze
Snowboarding	Women's slopestyle	Jamie Anderson	Gold
Snowboarding	Men's slopestyle	Sage Kotzenburg	Gold
Snowboarding	Women's halfpipe	Kaitlyn Farrington	Gold
Snowboarding	Women's halfpipe	Kelly Clark	Bronze
Snowboarding	Men's snowboard cross	Alex Deibold	Bronze
Speed skating	Men's 500 m relay	Alvarez/Celski/ Creveling/Malone	Silver
Ice hockey	Women's	Team	Silver
Figure skating	Ice dancing	Davis/White	Gold
Figure skating	Team event	Team	Bronze
Skeleton	Men's	Matthew Antoine	Bronze
Skeleton	Women's	Noelle Pikus-Pace	Silver
Luge	Women's singles	Erin Hamlin	Bronze
Bobsled	Two-woman	Meyers/Williams	Silver
Bobsled	Two-woman	Evans/Greubel	Bronze
Bobsled	Two-man	Holcomb/Langton	Bronze
Bobsled	Four-man	Fogt/Holcomb/ Langton/Tomasevicz	Bronze

In this activity we will use vertex colorings of graphs to determine optimal scheduling with regard to time under the following conditions:

- Interviews for medalists in sliding events will be held at the Sanki Sliding Center, while interviews for medalists of ice events will be held at the Iceberg Skating Palace. All other interviews of medalists in snow events will be conducted at the Rosa Khutor Alpine Center.
- Medalists who win on a two- or four-person team must be interviewed with their teammates and the four-person teams may be interviewed with other individual winners, but not another two- or four-person team.
- Due to space restrictions, the women's hockey team cannot be interviewed with any other medalists.

- Medalists in the same event should be interviewed in the same time slot.
- Further, the network wishes for the men's and women's freestyle skiers to be interviewed separately from each other, and the snowboarders and alpine skiers may not be interviewed together.
- Medalists in skeleton should not be interviewed with medalists in luge.
- The ice dancing figure skating champions do not want to be interviewed with the team event figure skating winners.
- And finally, in order to allow our audience to properly appreciate their achievements, gold medalists should not be interviewed with other gold medalists. (This excludes the team of ice dancers, who should be interviewed together.)

  1. Draw a graph to represent this scheduling problem.
  2. Color the graph using a minimum number of colors. Describe who is interviewed in each time slot. What is the minimum number of interview slots needed for each of the three locations?
  3. If you have a person in each of the three locations, what is the fewest number of time slots needed to complete all the interviews, assuming interviews in different locations can run concurrently? What if you have only one person conducting all the interviews: how many time slots are now needed?
  4. *Bonus*: Suppose each interview (except for the team hockey and team figure skating interviews) an only accommodate four people. How could you modify your schedule so the number of time slots is still minimal?

2.4 Projects

2.4.1 C. Froome

Team Sky

C. Froome, G.C. Rider

Manchester, England

Dear Graph Theory Student,

The sport of road cycling has had some bad press in recent years with the almost constant barrage of doping scandals. We at Team Sky, as a continuous sponsor and supporter of cycling, wish to explore ways to attract fans back to the sport. In particular, we would like to try to apply mathematics to optimize the year-long cycling experience.

As you know, the Union Cycliste Internationale, or UCI, is the governing body for international road cycling events such as the famous Tour de France or the Giro d'Italia. Much like in other sports, each year cycling teams compete in multiple events in order to accrue points which determine an overall winner at the end of the year. One such competition, called the UCI World Tour, is a year-long event consisting of 28 road races. Fourteen of these events are single-day races, including the five so-called Monument races, which are events of historical significance such as the Paris–Roubaix or Liège–Bastogne–Liège races and have been raced since the late 19th century. The other 14 events comprised 13 stage races (multiday races) and one team time trial.

Research the five Monument one-day races to determine the starting locations for each race. Create a directed weighted graph with race locations as the vertices, and directed edges representing the travel time or miles between all pairs of locations. For simplicity, your digraph needs only to include the starting location of each race and not the finish. Using a heuristic algorithm of your choice, try to find a minimum-weight Hamilton path. If you factor in travel times, one week of rest days between races, and the length of each race, how many days will it take to complete the five races? In addition to this data and graph, please compare your schedule with the traditional schedule used last year. Was an optimal schedule used? Explain.

We are optimistic that our joint effort will bolster the sport of cycling for the future.

Sincerely,
C. Froome

2.4.2 I. X. Ploor

I. X. Ploor
Freso, CA 93715

Dear Mathematically Inclined Travel Agent,

My family and I are very excited to continue planning for our upcoming travel. This year we have decided to do something different. We wish to use our two weeks of vacation to plan a week-long trip in December and another in June. We had already discussed a ski resort option, so I want to build upon this, and this year we want to explore the same place in both visits. In particular, we are looking to visit one of the large resorts in the United States, either the Killington Resort in Vermont or the Northstar California Resort, both of which you mentioned in previous correspondence. I believe that it will be a wonderful experience to ski down the slopes in the winter and then travel back to hike some of the same trails in the summer.

Choosing one of the resorts I mentioned above, I would like you to plan out an appropriate skiing and hiking route which will allow us to see all the resort has to offer. More specifically, you'll need to first find a current trail map and convert it into a digraph with trails directed down and ski lifts directed up. With the digraph, determine a route that will allow us to ski down every single trail and ride every lift in a fairly efficient way. My family are excellent skiers, so level of difficulty is no consideration, and, of course, the route should start and end somewhere at the bottom of the mountain. Then, because in the summer some of the trails are open for hiking, you will need to find a summer trail map and create a graph (no directions in this case) from these trails. To determine our hiking route, Eulerize the graph, finding a circuit of all trails beginning and ending at the bottom of the mountain.

I look forward to your help and my family's upcoming travel!

Regards,
I. X. Ploor

Implementation Notes The two resorts were chosen because they had easily available trail maps for winter and summer, but you may want to choose regionally appropriate resorts to appeal to your students. For ease of grading, you also might want to encourage the students to enlarge these and print them out in order to clearly draw the graphs directly on the maps.

2.4.3 C. A. Guzman

Comitê Olímpico do Brasil (COB)
C. A. Guzman, President
Rio de Janeiro, Brazil

Dear Cartography Expert,

In 2016, the Summer Olympics will be hosted by the city of Rio de Janeiro in my home country of Brazil. In the months preceding the games, the Olympic torch will travel throughout the major cities of the country. In order to promote community and inclusion, the torch will pass through each of the 26 states and the single federal district. In preparation for this massive event, we need to lay some new fiber optic cables.

Create a map showing the 26 states and the federal district of Brazil. Label the locations of the capitals of each state or district. Using these capitals as vertices, draw a dual graph to the state map. Applying an appropriate scale, approximate the distances by air between cities. Using an algorithm such as Prim's algorithm or Kruskal's algorithm, find a minimum-weight spanning tree which will determine the routes for the fiber optic cables connecting the 27 capital cities. Report the final approximate distance and time it will take to lay the cables.

We hope you are as excited as we are about these games, so we anxiously await your report.

Warmest regards,
C. A. Guzman, COB President

Implementation Notes The original version of this project included another task:

In addition to laying the cables, we need to determine a route for the Olympic torch. Use your map to describe a route through each of the 27 capital cities, keeping in mind that the final destination for the torch must be the Maracanã Stadium in Rio de Janeiro. Please provide the methodology for your choices of route. For example, are you choosing based on distance, time, topography, or some other reason?

However, this became a longer project than desired, so it is included here as an option.

3 Voting Theory

In this chapter, we present five activities and three projects that are concerned with voting theory. The concepts of voter manipulability and voting-system fairness are incorporated into some of the activities. The sports chosen to act as our representations for this chapter, baseball, basketball, football, and the Olympics, provide mostly context, and minimal sports knowledge is necessary to understand and complete the assignments.

3.1 Mathematics

Direct Systems

- Head-to-head—plurality, Condorcet, Dodgson
- Weighted—rank (point), Borda

Runoff Systems

- Hare
- Plurality runoff

Proportional Systems

- Approval voting

3.2 Sports

Baseball In Major League Baseball, the Cy Young award is given annually to the best pitcher from each league at the completion of the regular season. The award winner is determined by the Baseball Writers' Association of America, and over the years different voting systems have been used to determine the winners. The award is named after pitcher

Cy Young, who holds the record for the most career wins, with 511. (The pitcher with the second most career wins is Walter Johnson, with 417.) It was first awarded in 1956, the year after his death. To date, 17 Cy Young award winners have been chosen by a unanimous vote, and five of these were also awarded the Most Valuable Player (MVP) award for the best-performing player in the same year that they won the Cy Young award.[1] A project and an activity in this chapter will involve the Cy Young award.

Basketball Like Major League Baseball, the National Basketball Association (NBA) also awards a Most Valuable Player award, and one activity and one project in this chapter will be concerned with this award. The MVP award is given to the best-performing player at the conclusion of each regular season. Between 1956 and 1980 the winner was chosen by current NBA players; since 1981, however, the MVP has been selected by a panel of professional basketball writers and broadcasters. The winner of the award is given the Maurice Podoloff Trophy, which is named after the first commissioner of the NBA, who served from 1946 to 1963. Until recently, there had never been a unanimous winner of the MVP award, although Shaquille O'Neal for the 1999–2000 and LeBron James for the 2012–2013 seasons each fell one first-place vote shy. However, at the end of the 2015–2016 season Stephen Curry became the first unanimous inductee, garnering all 131 votes cast.[2] The second project related to basketball also concerns awards. Over the entire span of a career, the highest honor an NBA player can receive is to be inducted into the Naismith Memorial Basketball Hall of Fame (BHOF) in Springfield, Massachusetts. Named after the inventor of the modern game of basketball, James Naismith, the original Hall of Fame opened in 1959, has been relocated twice, and has seen multiple expansions. Today, over three hundred people have been inducted into the BHOF and the museum, located beside the Connecticut River, encompasses over 40,000 square feet. In addition to the historical exhibits, there are interactive components, including skills challenges, clinics, and shooting contests. The voting process for induction into the BHOF is secretive, with even the voters' identities kept secret.[3]

NCAA Football The process of ranking college football teams nationally has been a point of contention for decades and has often resulted in a split national championship, with two separate teams being declared national champion 11 times between 1954 and 2003. The NCAA has tried different methods for determining a best team from coaches' polls, media polls, computer polls, and combinations of these things, and currently uses a combination of a poll among a small group of experts and a bracket playoff system. Two activities in this chapter will be concerned with preseason and mid- or end-of-season rankings.

---

[1] Baseball Reference. "MLB Most Valuable Player MVP Awards & Cy Young Awards Winners." https://www.baseball-reference.com/awards/mvp_cya.shtml (accessed: November 28, 2018).

[2] Basketball Reference. "NBA MVP & ABA Most Valuable Player Award Winners." https://www.basketball-reference.com/awards/mvp.html (accessed: November 28, 2018).

[3] Basketball Hall of Fame. "History of the Naismith Memorial Basketball Hall of Fame." http://www.hoophall.com/history/ (accessed: November 28, 2018).

Olympic Games One activity and one project in this chapter will be concerned with how the International Olympic Committee (IOC) determines host cities for the summer and winter Olympic Games. No knowledge of any sports rules or awards is necessary to answer the questions.

3.3 Activities

3.3.1 2009 MLB Cy Young Award

Content Plurality system, Borda count, rank systems.

Activity The Cy Young award is determined by an election to determine the best pitcher in each league in baseball. The voters for the award are members of the Baseball Writers' Association of America, and in 2009 they were asked to rank the three best pitchers from each league. Recall that this type of ballot, where voters list a ranking of options, is called a *preference list ballot*. The total numbers of first-, second-, and third-place votes for the 2009 NL Cy Young award are given below in the table. Note that this is not a preference schedule, since we can't see how each individual voter ranked the pitchers.

Player	1st	2nd	3rd
Chris Carpenter	9	14	7
Dan Haren	0	0	1
Tim Lincecum	11	12	9
Javier Vazquez	0	1	0
Adam Wainwright	12	5	15

1. Determine who would win the election if the Cy Young award used a plurality voting system. What if a Borda count system was used to determine the winner? Winners for both of these systems can be computed from the above chart even though it is not a preference schedule.

2. Discuss the comparative pros and cons of plurality and Borda count voting systems in this context.

3. In fact, the Cy Young award election uses a *rank method* system and in 2009 each first-place vote was worth 5 points, each second-place vote 3 points, and each third-place vote 1 point. Using this rank method system, determine the number of points each pitcher would receive, and thus the Cy Young award.

4. Without seeing all 32 preference list ballots, it is impossible to know who would have been the Condorcet winner. Recreate a possible preference schedule to describe a possible collection of the 32 writers' votes knowing that two writers' ballots were as below:

Rank	(1)	(1)
1	Lincecum	Wainwright
2	Vazquez	Lincecum
3	Wainwright	Haren

5. From your recreated preference schedule, determine if there would have been a Condorcet winner for the 2009 National League Cy Young award.

Implementation Notes Questions 1–3 in this activity are fairly straightforward voting-theory computations, and students should be able to confidently answer them in small groups. However, groups often struggle with Question 4, as the notion of constructing a preference schedule from a vote total chart can be daunting. Expect to help students set up the preference schedule with the given first two columns, explain why an additional five or six columns will be necessary, and explain how to determine the possible distribution of voters for the columns.

3.3.2 NBA Most Valuable Player Award

Content Plurality system, Borda count, rank systems.

Activity The Most Valuable Player award is awarded to one player each year at the conclusion of the NBA regular season. The winner is determined by a vote where media members create a preference list ballot (a ranking) of the top five players they view as deserving. The ambiguity of what it means to be the most valuable player in the league is left to the voters to interpret; does it refer to the player who statistically has the most impact on a game, to who accounts for the largest proportion of his teams' points scored, or to some implicit value like who makes all of his teammates more valuable and play better? Some years there is a clear player deserving the award, and the voters' rankings, regardless of how the voters individually define "most valuable," easily reflect his position as the most valuable player, but in other years there has been great debate. Below we give the total numbers of first-, second-, third-, fourth-, and fifth-place votes for the top three players from the 2015 and 1990 MVP votes, representing both possible scenarios.

Player (2015)	1st	2nd	3rd	4th	5th	Player (1990)	1st	2nd	3rd	4th	5th
Stephen Curry	100	26	3	0	1	Magic Johnson	27	38	15	7	4
James Harden	25	87	13	4	0	Charles Barkley	38	15	16	14	7
LeBron James	5	12	62	32	12	Michael Jordan	21	25	30	8	5

1. Determine the plurality and Borda count winners for each of the 2015 and 1990 NBA MVP awards.

2. In fact, the winner of the award is determined by using a point-based system, with 10 points for a first-place vote, 7 for a second, 5 for a third, 3 for a fourth, and 1 for

a fifth. Knowing this information, determine how many points each player earned and who actually won the MVP award in each season.

3. Pretend that all the voters who had Michael Jordan first were forced to remove him from their ballot, moving Johnson and Barkley up to first and second on their ballots. Assuming that all of these voters originally had Magic Johnson and Charles Barkley ranked second or third, is it possible that Barkley would now be the winner of the MVP award? If not, how close could he come to winning?

4. What if we used a system based on powers of two, i.e., first place was worth twice as much as second place, second place twice as much as third, etc? Using such a system, assigning 1 point to a fifth-, 2 points to a fourth-, 4 points to a third-, 8 points to a second-, and 16 points to a first-place vote, would Stephen Curry and Magic Johnson still win?

5. In the 2015 MVP award vote, there were a total of 130 media members casting votes. Determine the percentage of the first-place votes that each player received. What percentage would be necessary to guarantee Curry would also be a Condorcet winner? Would you consider this to be a landslide win by Stephen Curry or a close win? (Keep in mind that the largest percentage a US presidential candidate ever received in the popular vote was in 1964, when Lyndon Johnson earned 61.05% of the popular vote.)

6. In the 1990 MVP award vote, there were 92 media members casting votes. Determine the percentage of first-place votes that each player received. Five times in US history, a presidential candidate has won the election while losing the popular vote: John Quincy Adams over Andrew Jackson (-10.4%), Rutherford Hayes over Samuel Tilden (-3.0%), Benjamin Harrison over Grover Cleveland (-0.8%), George W. Bush over Al Gore (-0.5%), and Donald Trump over Hillary Clinton (-2.1%).[4] How would Magic Johnson's first place vote deficit with Charles Barkley compare to these five presidential popular-vote deficits?

Implementation Notes Concerning sensitivities surrounding current events in the United States, faculty may want to consider removing the last question. Another option is to simply compare the Magic Johnson–Charles Barkley deficit with the largest one or two presidential-vote deficits.

3.3.3 Olympic Host City Decision Procedure

Content Plurality system, runoff systems, Condorcet system, Dodgson system.

Activity The host city for an Olympic Games is determined after a series of presentations and votes by nonvying members of the International Olympic Committee. In

---

[4] Encyclopaedia Britannica. "United States Presidential Election Results." https://www.britannica.com/topic/United-States-Presidential-Election-Results-1788863 (accessed: November 28, 2018).

2008, seven cities initially applied to host the 2016 Olympic Games: Baku (Azerbaijan), Chicago (US), Doha (Qatar), Madrid (Spain), Prague (Czech Republic), Rio de Janeiro (Brazil), and Tokyo (Japan). From these applicants, the IOC executive board allowed Chicago, Madrid, Rio, and Tokyo to go forward with their applications and become candidate host cities. Eventually, the cities each had ten official delegates from the host country make a 45 minute presentation to the IOC in October 2009. The presentations were immediately followed by questions from the IOC, and then a vote by secret ballot was held where IOC members from all countries not bidding to host cast a single vote for their top choice. The host city that received the fewest votes was removed as a candidate and a new vote was immediately taken. The process was repeated until a winning candidate was determined. The results are displayed in the table below.[5]

	Round 1	Round 2	Round 3
Chicago	18	–	–
Madrid	28	29	32
Rio de Janeiro	26	46	66
Tokyo	22	20	–

1. How can we tell that the IOC does not use an exact Hare system for this vote even though we see only each voter's first-place vote? Assuming that the IOC did in fact use a Hare system but that there was a recording error in the second round of voting, what are the likely possible vote totals that should have been recorded?

2. Assuming that voters who had Madrid or Rio as their first-place vote never changed their preferences, what information can we definitively infer about how the voters who initially voted for Chicago or Tokyo would have filled out a preference list ballot?

3. What information can we *not* discern about the voters' ballots?

4. Even though it appears likely, can we be sure there would be a Condorcet winner? If yes, explain why. If not, create a preference schedule that would agree with these votes totals (assuming that two Tokyo voters switched to Rio in Round 2) but would not yield a Condorcet winner.

5. Why do we know that Madrid could *not* be the Condorcet winner no matter how the preference list ballots looked? Would it be possible for Chicago, the first city eliminated, to have been the Condorcet winner? If so, create a preference schedule that would have the first-place votes agree with Round 1 and have Chicago as the Condorcet winner. If not, could Chicago have been a Dodgson winner?

[5] International Olympic Committee. "Election of the 2016 Host City." https://www.olympic.org/2016-host-city-election (accessed: June 22, 2018).

3.3.4 College Football Preseason Rankings

Content Rank method system, manipulability.

Activity The Amway Coaches Poll is conducted weekly throughout the regular season using a panel of head coaches at FBS schools. The panel is a stratified sample chosen by random draw, from each conference plus the independents, out of a pool of coaches who have indicated to the American Football Coaches Association their willingness to participate. Each coach submits a top 25 with a first-place vote worth 25 points, a second-place vote 24, and so on down to one point for the 25th place.[6] Notice that this is a rank-based method and is essentially a Borda count with over 100 options (teams) and 62 voters (randomly chosen coaches).

College football rankings have been a source of controversy concerning fairness in declaring a national champion. The NCAA has gone through multiple postseason structures, including a bowl system, the BCS, and a four-team playoff, to determine a champion. However you know very well that any election system with more than three options that includes a rank-based portion can be manipulated. A table showing the 2014 preseason rankings of the top 11 teams is below.

Rank	Team	Acronym	Points	No. of first-place votes
1	Florida State	FSU	1,543	56
2	Alabama	UA	1,455	0
3	Oklahoma	OU	1,382	3
4	Oregon	UO	1,314	1
5	Auburn	AU	1,271	0
6	Ohio State	OSU	1,267	1
7	California, LA	UCLA	1,085	0
8	Michigan State	MSU	1,050	0
9	South Carolina	USC	1,009	1
10	Baylor	BU	965	0
11	Stanford	SU	955	0

1. Suppose Coach X had his top 11 teams from 1 to 11 listed as

   FSU, UA, UO, AU, OSU, UCLA, MSU, USC, OU, SU, Missouri,

   and he felt *very* strongly that Oregon belonged in the top three. Assuming the other coaches' submitted ballots remain the same as in the table above and that Coach X has no knowledge of anyone else's submitted ballot, could he manipulate his ballot to bump Oregon into the number 3 spot? Explain why or why not.

---

[6] USA Today. "Amway Coaches Poll." http://www.usatoday.com/sports/ncaaf/polls/ (accessed: August 20, 2014). May not be available outside the US.

2. How many coaches who had the same "true" ballot would have to join Coach X in changing their ballots to put Oregon at 3? What would these coaches' disingenuous ballots look like, and what would be the new point totals for Oregon and Oklahoma?

3. Suppose Coach X also felt strongly that Stanford belonged in the top ten and not Baylor. Can Coach X manipulate his ballot to bump Baylor out of the top ten and Stanford into it?

Implementation Notes The notion of, and relationship between, secret ballots and single-voter manipulability is often difficult for students to understand. A conversation and role-playing vote concerning either events in class or local sports teams can help lead student groups toward a better start on this activity.

3.3.5 College Football In-Season Rankings

Content Rank method system, approval voting, manipulability.

Activity During the season, a select group of 13 people create a ranking of the best college football teams in the country known as the College Football Playoff ranking by blending approval voting with a rank-based system to create a top 25 ranking of the programs. This ranking is considered by many fans around the country as the official college football rankings, as it has the greatest effect on which teams compete in the National Championship four-team playoff and also how bowl bids are determined. The process for creating this ranking is as follows.

(A) Each committee member creates an unranked list of their top 30 teams. Teams listed by three or more members remain under consideration, while all other teams are discarded.

(B) Each member lists their top six teams from the remaining pool. The six teams included on the most lists comprise a subpool to now be considered.

(C) Each member now ranks these six teams and assigns 1 point to the first, 2 points to the second, etc. After each team's points are summed, the teams with the lowest three point totals are ranked 1, 2, and 3, respectively, while the three teams with the highest point totals are held over to the next round.

(D) Each member now lists their top six teams from the remaining pool (the original pool minus the six teams in the subpool). The three teams receiving the most votes are added to the other three unranked teams from the last round to create a new subpool.

(E) Steps (C) and (D) are now repeated until 25 teams have been ranked. This will always take seven rounds of voting (listing and ranking steps) to complete.

These rules and processes form the foundation for the voting process. There are some tie-breaking procedures as well, but we will generally stay away from such technical issues.

With these mid- and end-of-season ranking procedures in mind, consider the following questions.

1. In paragraph form, discuss the aspects of this voting system that mimic approval voting and those that mimic rank method voting. What benefits does the NCAA gain by having this blended system (i.e., how might it affect different fairness criteria)?

2. At what stages is it possible for a committee member to cast a disingenuous ballot to manipulate the ranking? Also, by doing such an action, how much can this member affect a team's ranking, assuming he or she is acting alone? This is quite important, especially for the original subpool of six, as only the top four teams are allowed to compete in the playoff for the national championship.

3. Is it possible for one of the originally chosen six teams in the first subpool to end up not ranked? If not, what is the lowest ranking guaranteed for those three teams in the first "holdover" group that make up half of the second subpool?

Implementation Notes This activity is best served as a follow-up to the previous one concerning college football preseason rankings. With experience working with vote manipulability, students can focus on how the following tiered voting system is susceptible. As the questions here lend themselves to writing, this activity can easily be made into an out-of-class assignment with students presenting their findings at the beginning of the next class period.

3.4 Projects

3.4.1 E. R. Axle

E.R.A. Pitching Academy
E.R. Axle, founder & owner
Norwell, MA 02061

Dear Mathematical Thinking Student,

My name is Emerald Rudolph Axle and I am the founder of the E.R.A. Pitching Academy in Norwell, Massachusetts. At my academy, you may have guessed that we train the future pitching stars for Major League Baseball. Historically, we have had more people applying for the academy than we could accommodate, but with the recent rise in popularity in other sports such as football, basketball, and juggling, we have had a significant decrease in business. I'm afraid that if we can't increase the number of applicants back to sustainable numbers, and quickly, I will have to close our doors forever. Now, you may be wondering why I've contacted you or how I found you, given the great distance between our homes. Well, you may remember previously working with my uncle, Ruby Insley, who runs a hitting academy in Hingham, and he gave me your contact information.

What I'm thinking is that we need to put together a new advertising campaign for the academy, and I've decided to focus on the Cy Young award winners from the American and National Leagues. The award has been around for a long time, since 1956, so lots of people have won the award over the years. To determine a winner, voters originally only picked their top choice and plurality voting was used. Then, in 1970, Major League Baseball instituted a rank-based system, but the exact nature changed periodically; sometimes voters chose only the top three pitchers (with a 5–3–1 scoring system) and other times they choose the top five pitchers and used a 7–4–3–2–1 scoring system. It certainly is confusing that the same award could be given out by different systems, and this got me to thinking how voting systems maybe have impacted the way the public evaluates pitching talent.

Now, I have plenty of employees that can produce the video aspects of the advertising campaign, but all of my best data miners are on vacation and this needs to be completed ASAP. Do you think you could create a report that would provide the necessary data for my video producers to create a convincing advertising campaign about the importance of pitching to success? I want to see the vote totals for the ten closest Cy Young award votes since 1956. In the years since 2012, it is very likely that you can find how each voter actually voted for the award. For these years, would it have made any difference if the 1970–2009 rank system using 5–3–1 or the plurality system of 1956–1969 was used to pick a winner? Does the Cy Young award appear to be fairly independent of or dependent on the voting system used to choose a winner?

A good place to look for some of the more recent data may be the Baseball Writers' Association of America website since those are the reporters who get to vote. Thank you so much, and I look forward to your report.

Sincerely,
E. R. Axle
E.R.A. Pitching Academy

3.4.2 H. Holmes

Searching for Truth
H. Holmes, Private Investigator
Farmington, ME 04938

Bloomsburg University Student,
You may be familiar with my distant relative Sherlock. He and I were both hired by the United States Olympic Committee (USOC) to dig into some irregularities with how the International Olympic Committee was awarding host cities for the Olympics and to advise them on how to better their chances of hosting an Olympics without resorting to bribery. Well, it turns out that the job is simply too outrageous for any one mind to comprehend, so we decided to focus on the corruption aspect of the job and hire some outside help for the advisement purposes. We hope that you will be able to analyze the information we have gathered to help advise the USOC on their next bid.

In particular, the USOC stressed that they are very impatient to submit a winning bid to host a Summer Olympics in the United States. To try to strengthen their application, we polled 100 retired Olympians about the five most important aspects of a host city from the athlete's perspective and asked them to rank their list. You can clearly see their opinions at the bottom of our letter. Now of course the USOC needs some say in which city will put forward the US bid, and they gave us the following cities as potential options: Charlotte, Denver, New Orleans, Philadelphia, San Diego, and Seattle. We need two things from you. First, collect all of the Olympians' opinions into a preference schedule (i.e., research these cities and, after determining how each group of the 100 Olympians would rank them, create a preference schedule of the cities). Then help us choose the winner that we should recommend to the USOC. We believe in the Condorcet fairness criteria, so please check that your method is fair, but would also like to know which cities are the Dodgson, plurality, Borda, and Hare winning cities. We will, of course, also need to see the raw data with references for your research.

   As always, your discretion is most appreciated, and we look forward to reading your speedy reply.

Sincerely,
Hemlock Holmes
P.I.

Key		Olympians' opinions					
Aspect	Abbreviation	(22)	(22)	(17)	(20)	(11)	(8)
Population	P	B	P	S	C	C	P
Safest	S	P	S	B	P	B	B
Beach access	B	C	C	C	S	S	C
Cleanest	C	S	B	P	B	P	S

3.4.3 L. Bird

Lucius Bird
Basketball Fanatic
Laramie, WY 82070

Dear Mathematical Thinking Student,

   I am writing you in the direst of circumstances and need your help. My name is Lucius Ervin Bird and I am the *biggest* basketball fan under 4' 10" as I stopped growing at age 12 due to a bad egg salad sandwich. Living in Laramie, we don't have an NBA team, so I watch whichever game I can get on my old television. Many fans may think this is unfortunate, since I don't have a "hometown" team to root for, but in fact it's a great advantage as I can appreciate the sport for its beauty and the necessary skills to play well rather than being influenced by geographical biases.

   This led me to my current predicament. The voting procedure for induction into the Hall of Fame is shrouded in secrecy, and even the *biggest* NBA fan like myself can't analyze prospective players' probabilities of enshrinement. First, let me explain what is known about the process before I delve into my task for you. Players, coaches, referees, and otherwise influential basketball pioneers are eligible for enshrinement so long as they are retired for a minimum of 5 years, at which point a nomination packet must be submitted to the Naismith Memorial Basketball Hall of Fame (BHOF). A screening committee reviews the packets and puts forward a maximum of ten applications for the Honors Committee to consider. The Honors Committee is composed of 24 members and a candidate needs 18 votes to be enshrined in the BHOF.

And that's all we know! We don't know how candidates vote, whether they submit a preference list ballot with rankings or an approval ballot without rankings, what constitutes a vote for a candidate (only a first-place ranking, top three, etc.), whether or not there is any discussion and multiple rounds of voting, whether there is a maximum number of enshrinements per year, or any other relevant information. My thoughts are these: if we can write a report discussing one or two viable voting options and highlighting how these processes are fair for election to the hall, then maybe the BHOF will finally feel comfortable to publicize their process. The trick is to guess which voting procedure they use so that our report is applicable. I want you to write a report discussing how approval voting and rank-based voting could be used to elect enshrinement into the BHOF. You should discuss how different cutoffs for the approval voting or point totals could affect candidacies and also whether instituting a minimum or maximum number of enshrined members would be worthwhile.

Thank you so much for your help, and I hope to enjoy reading your report on my next camping trip out under the great open sky of the west.

Sincerely,
Lucius Bird

Implementation Notes Sometimes students do not have a solid understanding of a cutoff in these circumstances. It may be beneficial to have them take a few minutes to discuss how they might approach this part of the assignment in small groups, before the students attempt to work individually.

4 Fair Division and Apportionment

The activities and projects in this chapter apply strategies of fair division and apportionment as well as investigate the use or occurrence of gerrymandering within the confines of soccer, basketball, and football.

4.1 Mathematics

Fair Division

- Divider-chooser
- Lone divider
- Last diminisher
- Sealed bid
- Moving knife

Apportionment

- Hill–Huntington method
- Hamilton's method
- Jefferson's method
- Webster's method
- Gerrymandering

4.2 Sports

Basketball The National Basketball Association is the organization governing professional basketball. Professional basketball is known for its superstars such as Michael

Jordan and LeBron James. Because each team has only five players on the court at one time and at most most 15 players on a roster, individual talent is more apparent and exceptional players can make a big difference. This is unlike professional football, which puts 11 players on the field at a time and can have a roster of over 50 players. Keeping the superstars and other players happy and working well with each other is a big part of an NBA coach's job. In this chapter, the basketball activity uses an apportionment strategy to optimize playing time for a professional basketball team.

Football Both the National Football League (NFL) and college football provide the backdrop for activities in this chapter. Football is the most popular sport in the United States. In fact, if you consider the NFL and college football as individual sports, each is more watched than any other sport. Two activities and a project in the chapter are more focused on the business of football. The first activity applies the sealed bid method to fantasy football trades. The second activity explores the division of the country by college football recruiters. A project also deals with the NFL, considering the current and potential divisions of professional football teams.

Soccer Two activities and two projects in this chapter are set in the context of soccer, also known as association football. Soccer's governing body is the Fédération Internationale de Football Association (FIFA). FIFA is responsible for organizing all the qualifying matches in the four years leading up to its premier event, the World Cup. Host sites for the World Cup, FIFA's most well-known tournament, which occurs every four years, are apportioned in the first activity. The second activity focuses on video game rankings of specific skills for the top players in the world and how to apportion advertising time highlighting these players. The first of the two projects investigates how an unscrupulous tournament official could distribute teams in the World Cup for his own benefit, and the second visits a small soccer club needing to divide incentives fairly among players.

4.3 Activities

4.3.1 FIFA World Cup Host Countries

Content Hamilton, Jefferson, and Webster's apportionment methods.

Activity Since the first World Cup in 1930, countries have contended for the honor of hosting the tournament. Given the great disparity in travel distance for teams depending on the location, fair allocation of sites is necessary. After early boycotts because of the choice of location, FIFA typically alternated the sites yearly between Europe and North and South America.

A July 2017 ranking of the top world soccer national teams is as follows:[1]

1. Germany	11. Spain	21. Slovakia	31. Ecuador	41. Paraguay
2. Brazil	12. Italy	22. Northern Ireland	32. Netherlands	42. Serbia
3. Argentina	13. England	23. Iran	33. Turkey	43. Romania
4. Portugal	14. Peru	24. Egypt	34. Tunisia	44. Burkina Faso
5. Switzerland	15. Croatia	25. Ukraine	35. USA	45. Australia
6. Poland	16. Mexico	26. Costa Rica	36. Cameroon	46. Japan
7. Chile	17. Uruguay	27. Senegal	37. Austria	47. Denmark
8. Colombia	18. Sweden	28. Congo DR	38. Greece	48. Algeria
9. France	19. Iceland	29. Republic of Ireland	39. Nigeria	49. Haiti
10. Belgium	20. Wales	30. Bosnia and Herzegovina	40. Czech Republic	50. Ghana

Categorize the countries by region: North America, South America, Europe, Africa, and Asia/Australia.

1. Use Hamilton's method to apportion the next 20 tournaments to these five regions.
2. Use Jefferson's method to apportion the next 20 tournaments to these five regions.
3. Use Webster's method to apportion the next 20 tournaments to these five regions.
4. Were your allocations the same? Which method do you prefer? Explain.

Bonus: A 2014 article[2] ranks the fanbases of teams that competed in the 2014 World Cup, held throughout Brazil. Use a device to find this article. What if you apportioned the next 20 tournaments based on these teams instead of the top 50? Use your preferred method to complete the allocation. Did the results change? Explain why or why not.

Implementation Notes Categorizing the countries will probably cause the students some difficulty. Since a correct categorization does not really affect the mathematical content, you may wish to be flexible in the categorization as long as the students are consistent among the different methods. This shortens the amount of time spent Googling countries at the beginning of the activity.

---

[1] FIFA. "Men's Ranking." http://www.fifa.com/fifa-world-ranking/ranking-table/men/index.html (accessed: November 29, 2018).

[2] McNicholas, James. "Ranking All 32 Nations' Fanbases at 2014 World Cup." Bleacher Report. June 7, 2014. http://bleacherreport.com/articles/2085627-ranking-all-32-world-cup-fanbases (accessed: November 29, 2018).

4.3.2 World Cup Advertising

Content Hill–Huntington apportionment method.

Activity For their video game, EA Sports[3] ranked the top six nongoalie soccer players in the world as Ronaldo, Messi, Neymar, Suarez, Lewandowski and Ramos. The rankings were calculated using pace, shooting, passing, dribbling, defending, and physicality as given in the table below.

Rank	Name	Pace	Shooting	Passing	Dribbling	Defending	Physicality
1	Ronaldo	90	93	82	90	33	80
2	Messi	89	90	86	95	26	61
3	Neymar	92	84	79	94	30	60
4	Suarez	82	90	79	86	42	81
5	Lewandowski	81	88	75	86	38	82
6	Ramos	76	63	71	72	88	83

Suppose FIFA wants to created a special marketing campaign that will highlight these players and hopefully draw additional interest in the competition (not that the World Cup needs it). The marketing campaign will be composed of eight segments, and each segment will focus on one individual player from the above list and his country of origin. Players may be the focus of multiple segments but each segment will be on exactly one player.

1. Given the above point totals for individual skills, assign a value to each player using the following formula:

$$V = 100*\text{Defending} + 10*(\text{Pace} + \text{Physicality}) + 1*(\text{Shooting} + \text{Passing} + \text{Dribbling}).$$

2. Using this value in place of votes, determine the qualification value for the Hill–Huntington method. Are any players eliminated from the marketing campaign?

3. Using Hill–Huntington, determine the allocation of segments to players.

4. For a comparison, if segments *could* have focused on multiple players (i.e., allow for fractional segments), what proportion of segments does each player deserve?

5. Suppose all players must be allocated at least one full segment; repeat the above Hill–Huntington calculation without disqualifying any players, and initially assigning one full segment to each player.

[3] EA Sports. "FIFA 18—Player Ratings Top 100." https://www.easports.com/fifa/fifa-18-player-ratings-top-100 (accessed: November 29, 2018).

Implementation Notes Due to the nature of the Hill–Huntington method, this activity is one of the longer activities and is calculation heavy; we suggest, in the prior class, asking students to bring calculators or laptops to the classroom to make it go quicker. Additionally, subsequent Hill–Huntington apportionments can be completed by using different weighting systems or eliminating some of the skill categories. Out-of-class assignments can include the requirement that students initially find rankings for the top players in their favorite sports before allocating the marketing segments via Hill–Huntington and comparing the results with other apportionment methods.

4.3.3 Playing Time

Content Apportionment methods.

Activity Coach Lue has decided to determine a player's playing time based on average points scored during playoffs. With a total average of 127.4 points per game, the distribution of average points per player for the 2017 playoffs is as follows.[4]

Player	Frye	Irving	James	Jefferson	D. Jones	J. Jones	Korver
Avg points	12.8	25.9	32.8	3.9	1.6	0.2	5.8
Player	Love	Shumpert	Smith	Thompson	Dn. Williams	Dk. Williams	
Avg Points	16.8	4.4	8.1	8.2	4.3	2.6	

We wish to use these statistics to allocate minutes per game.

1. In order to keep all players happy, every player must play a minimum of 12 minutes per game. Use your favorite apportionment method to allocate playing time in a 48 minute game. Recall that five players play at a time, so this is equivalent to allocating 240 minutes of playing time where 156 minutes are automatically allocated by the restriction.

2. Suppose the coach decides to allocate playing time based on the match-up with the opposing team. This week he expects the game to be close, so only players scoring an average of 5 points or higher will play in the game.

   (a) Use your apportionment method to allocate playing time among the seven players whose average points were 5 or higher. Do you notice a problem?

   (b) Modify your apportionment method to cap the minutes played at 48 minutes a game for an individual player. Hint: If a player was allocated above 48 minutes in part (a), set his minutes to 48, remove him from the calculation, I and redo the apportionment.

[4] Real GM. "Cleveland Cavaliers Playoff Roster." https://basketball.realgm.com/nba/teams/Cleveland-Cavaliers/5/Rosters/Playoff/2017 (accessed: November 29, 2018).

3. Would you want to use a modified method such as you used in Question 2, with an upper limit on the number of representatives, to apportion a house of representatives? Can you think of another situation where such an upper limit is appropriate?

Implementation Notes There are many variations to this type of activity depending on the strength of your class. Students could investigate other statistics or combinations of statistics to create a metric which could be used to apportion playing time. More statistically minded classes could test the correlation of apportioned playing-times from a particular statistic with actual playing time data to determine if any are good predictors.

4.3.4 Fantasy Free-Agents

Content Sealed bid method.

Activity In a fantasy football league, six high-performing players are free agents and are being competed over by four team owners. The teams decide to work together for optimal happiness, hoping to improve against other teams in the league. The players and the salaries each team owner is willing to pay are given in the following table.

| | Players | | | | | |
	T. Gurley	R. Gronkowski	O. Beckham	A. Brown	K. Hunt	D. Watson
Game of Jones	5,500	8,000	4,000	6,000	6,500	9,500
Le'Veon la Vida Loca	7,500	6,000	3,000	9,000	8,500	4,500
The Walking Dez	9,000	4,000	4,500	4,000	8,000	7,000
Yo Belichick Yo Self	5,500	9,000	5,000	4,500	7,500	9,000

1. Use the sealed bid method to allocate these six players. Determine which team gets each player and calculate each team's net payout or compensation.

Before the bids are placed, Le'Veon la Vida Loca worries that she will not have enough cash to cover her bids, so she makes some changes.

	T. Gurley	R. Gronkowski	O. Beckham	A. Brown	K. Hunt	D. Watson
Le'Veon la Vida Loca	7,000	5,500	3,000	9,000	7,500	4,500

2. Use the sealed bid method to allocate the players and calculate each team's net payout or compensation.

3. What if we switch to the player's perspective? Suppose the players bid salaries that they would be willing to earn from each team. Could you use a fair division method to allocate each player to a team he would be content to play for? Explain.

Implementation Notes We acknowledge that in traditional online fantasy football formats these kinds of deals are not usually made, but we ask the students to play along all the same.

4.3.5 College Football Recruiting

Content Lone divider, moving knife, and last-diminisher methods.

Activity Four recruiters for the University of Kentucky need to divide the country into recruiting areas. The head coach divides the country into regions labeled Midwest, West, East Coast, and South. Each of his recruiters secretly assigns a proportion to each of the four regions that indicates his or her perception of the value of that region. This information is shown below.

Recruiter	Midwest	West	East Coast	South
Woody	35%	10%	25%	30%
Bear	20%	15%	45%	20%
Knute	40%	10%	10%	40%
Frank	40%	10%	25%	25%

1. For each recruiter, list any regions at he or she might feel have a fair share.
2. Can you allocate the regions so that each recruiter is happy?

Woody grabs the map and proceeds to divide the country into new regions, Top, Bottom, Left, and Right, which he thinks divide the country fairly. The other recruiters now secretly assign proportions for their preferences for the new regions as follows:

Recruiter	Top	Bottom	Left	Right
Woody				
Bear	40%	25%	15%	20%
Knute	30%	30%	30%	10%
Frank	25%	35%	20%	20%

3. What proportions did Woody assign to each region?
4. Can you allocate the regions so that each recruiter is happy? What is this allocation?

5. Why did the second allocation work, but not the first? What method was used in this second allocation?

6. Would either the last-diminisher or the moving knife method work in this problem? Explain how these methods would work in this situation.

7. Can you describe another method of fair division that might have worked on the first scenario where the coach divided the country?

Implementation Notes This activity emphasizes that the lone divide method is not guaranteed to work when the divide is not one of the parties with a stake in the outcome.

4.4 Projects

4.4.1 Connor Tist

Connor Tist
P.O. Box 4567
Zurich, 8001, CH04

Dear Football Analyst,

Thank you for your willingness to help me solve this problem. As you know, after the qualifiers are completed the World Cup field consists of 32 teams. For the 2018 World Cup, those teams included five teams from Africa, five teams from Asia, 14 teams from Europe, three teams from North and Central America, and five teams from South America.[5] First, I'll need you to rank these teams based on their all-time FIFA World Cup points or by current world ranking. Be sure to describe the ranking method used and give me a reference to your source material.

In the initial stage of the World Cup, the teams are placed into eight groups of four teams. In 2018, the best eight teams, seeded 1–8, we are placed in the first pot, the teams seeded 9–16 were placed in the second pot, the teams 17–24 in the third pot, and the teams 25–32 in the fourth pot. Via a random draw, one team from each pot was placed in each of the eight groups. This method replaced the process from previous years, where the top eight seeds were placed in one pot and the remaining 24 teams in another pot. The top seeds were placed in separate groups and the rest of the teams were randomly assigned to a group (with some restrictions on geography, which we will ignore here). As you will see, part of the justification for such a change was to deter unscrupulous activities. After the draw, the group round begins and the winners of each group then compete in a single-elimination tournament to determine the champion.

Using the method prior to 2018, suppose some unscrupulous FIFA official (no one I know, of course) has placed a bet that at least three European teams will advance to the single-elimination tournament. Barring an upset, can he choose the eight groups to ensure that this happens? Please provide his proposed allocation and explain his

---

[5] FIFA. "2014 FIFA World Cup Brazil." https://www.fifa.com/worldcup/archive/brazil2014/teams/ (accessed: November 29, 2018).

strategy. What if the restriction on seeding the top eight teams is removed? How would this change his decision?

Suppose instead, still hypothetically of course, he bets that two European teams and one South American team will advance out of the group rounds. How should he choose the eight groups to ensure the greatest chance of winning his bet? Can he guarantee that this will happen?

Finally, how could you place the teams to give the best chance to win a long-shot bet on a North American winner? I appreciate any information on this truly imaginative situation and I'll need the results as soon as possible, of course.

Sincerely,
Connor Tist

4.4.2 D. Roepkik

Coach D. Roepkik
Citrus Heights Lemons
Citrus Heights, CA 95611

Dear Fair Division Specialist,

As a coach and owner of a newly formed and not too highly funded soccer club, I don't have enough cash to pay all the player salaries. After raising enough to pay my top seven players, four are left without pay. Getting creative, I have visited local businesses and received sponsorship items which I'll be able to use to compensate the players. I met with each player and asked them to bid on each item, with the following outcomes.

Prize	Players			
	1	2	3	4
55 inch flat screen TV	500	450	600	375
Dinner for two at Applebees	25	34	49	31
Used Play Station 4	124	200	159	170
Year's supply of car washes (one per week)	310	227	150	99
County fishing license	32	5	18	10
5 hours of babysitting	0	25	29	40
One bucket of used golf balls	64	75	100	23
10 single-topping pizzas	35	58	77	65
Tax filing services (state return not included)	100	120	150	99
New iPhone 6	50	200	180	175

Please help me divide these resources so that each player is satisfied. Name and describe the method you are using. I understand there may be some cash payments by players. In your report, please specifically explain how each of these payments is calculated and why this is fair, so that I can convince the players that this method truly works.

Finally, for tax purposes, I need to know the actual market value of these ten items. Research typical costs and include these costs and your references in the report. Based on this data, rank the players by total compensation in terms of actual market value.

Thanks for your help getting the Citrus Heights Lemons off to a great year!

Sincerely,
Coach R.

4.4.3 R. Goodell

R. Goodell, Commissioner
National Football League
New York, NY 10154

Dear Gerrymandering Expert,

The National Football League is made up of two conferences, the American Football Conference (AFC) and the National Football Conference (NFC). Each of these conferences is separated into four divisions, which are named for regions of the country, specifically, East, North, South, and West. While these divisions may have originated in a reasonable fashion, because of expansion and team movement many of these divisions do not make very much sense geographically. First, I need you to create a graphic of the United States divided into eight regions illustrating potential districts for these eight divisions. In order to argue for realignment, I'll need you to explain to me which of these regions make sense geographically and which do not.

Now, supposing the current 32 teams remain in the same cities, how would you propose realignment, that is, changing the teams within the divisions, so there is less travel and an opportunity for more vigorous rivalries? Keep in mind that each division must still have four teams and that current AFC and NFC teams must remain in AFC and NFC divisions, respectively. Further, you should strive to move the fewest teams in order to minimize upheaval in the league. Create a new illustration of these eight divisions.

Finally, there are yearly rumors that the NFL is looking to expand. If the league decided to add four teams, two in the AFC and two in the NFC, we wolud need to allocate these teams to existing divisions. Potentially rumored cities are the international locations of Toronto, London, and Mexico City, and as well there is the possibility of adding another team to Texas in San Antonio. If teams were located in each of these cities, to which divisions would you add these teams? Consider where these teams would be placed in both the current divisional setup and your modified divisions.

I thank you for your efforts to improve the great game of professional football.

Sincerely,
R. Goodell, Commissioner

Implementation Notes To allow even better geographical alignment, another variation is to relax the requirements on the conferences and let the students place any set of teams in eight distinct divisions.

5 Financial Mathematics

In this chapter, we present five activities and three projects that are concerned with percentages, saving and borrowing models, and stocks. The sports chosen as a medium for this material are basketball, in particular the NBA, and the Olympics. As the material deals with only the financial aspects of these two sports, no knowledge of any sports is necessary.

5.1 Mathematics

Percentages

- Cost overruns
- Interest versus principal

Saving and Borrowing Models

- Simple and compound interest
- Present and future value
- Mortgage rates
- Cost-of-Living Index
- Amortization

Tax Rates

- Federal and state income taxes
- Sales taxes

Stocks and Stock Indices

- Dow Jones
- S&P 500

5.2 Sports

Olympic Games One activity and two projects in this chapter are concerned with financial decisions that planners, cities, and nations must make concerning cost and financing related to the Olympic Games. Another activity explores the financial repercussions for fans who take out short-term loans to attend the Olympics.

Basketball Three activities and one project in this chapter will be concerned with financial decisions individual players make concerning their contracts. The first activity is concerned with purchasing power and home mortgage rates. The second activity in this chapter is concerned with the bonus money, referred to as the "playoff bonus pool," that the NBA distributes to teams for reaching and advancing through the playoffs. Each year the total amount of the playoff bonus pool is increased, so comparing earnings across long time spans can be misleading. Each team is allowed to divide the money however they feel is most fair, but most teams divide the money equally among all players and staff on the team during that year. There are some discrepancies concerning whether there should be a minimum number of games played for a player to receive a share and whether the shares should be proportional to salaries, but for our activity, we assume equally sized shares and no minimum-games-played requirement. The third activity and one project in this chapter are concerned with the present value of a maximum contract. Each year, the NBA sets a limit for the maximum number of dollars a team is allowed to pay any one player signing a new contract. This maximum amount is typically called a maximum contract and is in effect on a yearly basis. However, although the gross income of a player signing a maximum contract is constant regardless of where he plays, other factors affect the net income, including taxes and cost of living. In addition, some players will sign multiyear contracts, agreeing to play in a city for two, three, or more years at set salaries. In this multiyear scenario, concerns related to present versus future value and investment opportunities also must be considered when determining the net worth of a contract.

Implementation Notes The material covered by these activities is very computation heavy. With this in mind, we recommend telling students via email or in a previous class to bring either a calculator or a laptop to class for the dates on which you plan to utilize one of these activities.

5.3 Activities

5.3.1 Olympic Host Costs

Content Percentages.

Activity The Olympic Games can be an extremely costly affair for the host nation and city. Estimated costs are often exceeded, sometimes vastly, leaving the local population to make up the difference. The table below shows the total costs and percentage cost overruns of the three summer Olympic Games between 2004 and 2012.[1]

Games	Country	Type	Final cost, billion USD	% Cost overrun
London 2012	UK	Summer	14.8	133
Beijing 2008	China	Summer	5.5	35
Athens 2004	Greece	Summer	3.0	97

1. The Boston proposal to host the 2024 summer Olympic Games estimated a total cost of $11.8 billion.[2] Determine the overrun and total cost of the 2024 Olympic Games assuming each of the different possible percentage cost overruns from the 2004 through 2012 Olympic Games.

2. The Massachusetts GDP for 2015 was $427,541,000.[3] What percentage of the state's GDP would each of the above overrun costs from Question 1 be?

3. If the state of Massachusetts was responsible for the entirety of the overrun costs, it would have many options for financing the cost, and here are a few standard ones:

   (a) Issue state bonds valued at $1000 with 10 year maturity, sold at 1.63% simple interest.

   (b) Raise taxes such as the gasoline or cigarette tax. A flat tax on such products could be a 10¢ tax on every gallon of gas or pack of cigarettes.

   (c) Decrease funding to state-sponsored programs such as in-state tuition discounts at state universities. Such a decrease could be seen as the 2016 tuition and fees being raised from $14600 (about 42% of the out-of-state tuition) to $20000 (about 64%).

   If the state of Massachusetts instituted any of these options, how many bonds, gallons of gas, or in-state tuitions would be needed to make up the costs?

4. Knowing that in 2016 the state had a population of about 6.8 million residents, is any one of these options a realistic fix to make up the overrun costs? What if the state instituted all three?

[1] Flyvbjerg, Bent and Steward, Allison. "Olympic Proportions: Cost and Cost Overrun at the Olympics 1960–2012." Saïd Business School Working Papers, Oxford: University of Oxford. June 1, 2012. 23 pp. http://dx.doi.org/10.2139/ssrn.2238053 (accessed: July 20, 2016).

[2] Bazelon, Coleman et al. "Analysis of the Boston 2024 Proposed Summer Olympic Plans." The Brattle Group. http://www.mass.gov/governor/docs/news/final-brattle-report-08-17-2015.pdf (accessed: July 20, 2016) page 28.

[3] Department of Numbers. "Massachusetts GDP." http://www.deptofnumbers.com/gdp/massachusetts/ (accessed: July 1, 2016).

5. The expected overrun cost for just the Olympic Stadium was $83.6 million.[4] Would this be a feasible cost to cover with any one of the options?

5.3.2 Olympic Attendance Cost

Content Percentages, amortization, credit cards.

Activity The 2018 Winter Olympics were held in Pyeonchang, South Korea, and hundreds of thousands of fans from the 90 countries sending athletes attended the competition from February 9 to February 26. As it was a once-in-a-lifetime opportunity, many students decided to take some time off school to attend. Aside from the missed time at university, students needed to determine a manageable way to defray the costs, as a sample budget including airfare, meals, tickets, and hotel accommodation could approach $5,000 ($1,000 for the flight and $4,000 for everything in Korea),[5] so many used a credit card to allow them to make payments over time. Below are the details for student credit cards from three major credit card companies.

Credit card company	Card name	APR (%)	Cash back	Annual fee($)
Discover	Student Chrome Card	23.49	2% for first $1,000 quarterly, 1% afterwards	0
Mastercard	Deserve Edu	20.24	1.00%	0
Visa	Journey Student Rewards	24.99	1.25%	0

1. Considering each of the cash-back benefits, what is the total balance due after a student charges the $5,000 to cover the costs of attending the Olympics?

2. Calculate the standard monthly payment for each card if the student wanted to pay off the credit card after a total of 5 years. For each card, what is the total amount the student will end up paying for the $5,000 initial charge? Use the amortization formula for monthly payments given below:

$$A = \frac{P * (r/12)}{1 - (1 + r/12)^{-12t}}$$

---

[4] Bazelon, Coleman et al. "Analysis of the Boston 2024 Proposed Summer Olympic Plans." The Brattle Group. http://www.mass.gov/governor/docs/news/final-brattle-report-08-17-2015.pdf (accessed: July 1, 2016) page 65.

[5] Leonhardt, Megan. "How Much a Trip to the 2018 Winter Olympics Would Cost You." Time. November 9, 2017. http://time.com/money/5013664/winter-olympics-2018-ticket-prices-costs/ (accessed: May 15, 2018).

3. What if the student wants to pay off the charge over 10 years? What are the monthly payment and total cost?

4. Minimum payments vary by card but can be approximated by taking 2% of the balance due for balances over $1,000$ and simply $35 for balances under $1,000$. Calculate the minimum payment for each of the above cards.

5. Using the amortization formula above, how long will it take the student to pay off each of the credit cards if she makes only the original minimum payment each month? What is the total cost for the student in this scenario for each card?

6. Additionally, the cards offer some benefits that target certain students: the Discover pays $20 to account members annually that maintain a 3.0 GPA; the Mastercard pays for a 1 year Amazon Prime Student Membership ($49); and the Visa has no foreign transaction fees. (The other two in fact have this fee; it is normally a 3% fee for all charges made in another country.) Do you expect these amounts to make a difference to which card is the best deal for the student? Discuss what assumptions would need to be made and how one would calculate the total cost for each card with this new information.

Implementation Notes It is a good idea to remind students that credit cards actually have interest compounded daily. However, we found the calculation became increasingly cumbersome using this in the activity.

5.3.3 Home Mortgage Rates

Content Amortization, compound interest, mortgage rates.

Activity In the summer of 2016, Al Horford was an NBA free agent and signed a 4 year contract to play with the Boston Celtics. Coming from Atlanta, where he earned a total of $60,000,000$ over 5 years playing for the Atlanta Hawks, he needed to purchase a home for his time in Boston. Boston is an expensive city to live in; a nice four-bedroom condo in Beacon Hill can cost $5,000,000$ or more. For practical purposes, let us suppose the house to be considered is priced at $5,000,000$. Mortgage rates for July 1, 2016 were approximately 3.375% for a 30 year loan, 3.250% for a 20 year loan, and 2.750% for a 10 year loan. Recall that the formula to determine the regular monthly payment, $\mathcal{R}$, of a loan can be given as:

$$\mathcal{R} = \frac{P * (r/12)}{1 - (12/(12+r))^{12t}}$$

where P is the loan amount, r is the annual interest rate, and t is the number of years needed to repay the loan.

1. Calculate the monthly payment for the 30 year, 20 year, and 10 year loans.

2. What would be the total amount paid if Al Horford paid off each loan in the maximum amount of time?

3. What percentage of each total amount paid was interest?

4. After 1 year, can we determine how much principal is owed on each loan? Notice that, for each month,

$$\text{Monthly Payment} = \text{Interest Payment} + \text{Principal Payment}.$$

To determine an interest payment, we can simply calculate

$$\text{Loan Balance} \times r/12.$$

Create a loan amortization chart for the first year of each of the 30 year, 20 year, and 10 year loans. Keep in mind that the loan balance changes each month, as you must subtract each principal payment.

5. Assuming he decided to stay in Boston after his contract ended, and continued to pay off his mortgage, how long would it take to pay off each type of mortgage? What would be his total amount paid and his total interest paid, and what percentage of his total loan payment would be interest?

Implementation Notes The following bonus problem can be computationally cumbersome for students without proper direction. Be sure to remind them to use the amortization chart that they created in Question 4 as an aid.

6. *Bonus.* Suppose Al Horford had a savings account of $7,000,000 that earned 0.05% interest, compounded monthly, that he used for housing expenses in the following manner. On June 30 of every fifth year, he used all accrued interest to make a payment against the principal of his current mortgage. Estimate how long it would take him to pay off the 30 year loan and how much he would save on interest by making these extra payments.

5.3.4 Playoff Bonus Pool 2011 vs. 2016

Content Tax rates, compound interest, indexed stock returns.

Activity The NBA pays players and teams additional money, called the "playoff bonus pool," to their yearly salary for making and progressing through the playoffs. Each team divides their portion of the pool into equal payments given to all players, coaches, and staff on the team that year. For players on teams that make the playoffs consistently, this bonus money can be a significant source of income. We are going to compare the bonus money that one superstar in the NBA, LeBron James, earned in the 2011 season when he was on the Miami Heat with what he earned in the 2016 season when he was on the Cleveland Cavaliers. The question we wish to answer is as follows: In terms of value in July 2016, is the bonus money for winning the 2016 NBA championship worth more or less than coming in second in 2011?

1. The 2011 Miami Heat had 17 players and 24 coaches and staff on the roster, while the 2016 Cleveland Cavaliers had 18 players and 25 coaches and staff. Using the data in Tables A.1 and A.2 and assuming that each person on the team was given an equal share of the playoff bonus pool for a given year, determine LeBron James' share.

2. The maximum federal income tax for each year was 39.6%. The maximum state income tax for Ohio in 2016 was 4.997%, while Florida did not have a state income tax in 2011. Determine the net pay that LeBron James earned after paying the appropriate federal and state income taxes on the bonus. Does the fact that Florida does not have an income tax make up for the smaller gross payment compared with winning the championship with Cleveland?

3. On July 1, 2011, the rate for a 5 year certrificate of deposit (CD) was 1.61%. If the full 2011 net share was invested in such a CD on July 1, 2011, what would it have been worth on July 1, 2016?

4. Between July 1, 2011 and July 1, 2016, the Dow Jones returned 62.430% on an investment assuming all dividends were reinvested. If the full 2011 net share was invested in the Dow Jones on July 1, 2011, what would it have been worth on July 1, 2016? If you were to use the average annual growth of the Dow Jones found in Table A.3 rather than the 4 year return, what would it be worth? Assume the market compounds daily.

5. Between July 1, 2011 and July 1, 2016, the S&P 500 returned 75.485% on an investment assuming all dividends were reinvested. If the full 2011 net share was invested in the S&P 500 on July 1, 2011, what would it have been worth on July 1, 2016? If you were to use the average annual growth of the S&P 500 found in Table A.4 rather than the 4 year return, what would it be worth? Again, assume the market compounds daily.

6. Assuming that federal income tax had to be paid on the interest and/or gains of each of these investments, what would be the net worth in 2016 of investing the 2011 playoff bonus pool share in a CD, the Dow Jones, or the S&P 500? Complete this problem using both the 4 year return and the average annual growths, removing federal taxes each year.

Implementation Notes The data in Tables A.1 and A.2 contains information on all the teams' bonus money in the years 2011 and 2016, so the activity may be easily modified for teams in your region.

5.3.5 Maximum Contract 2016

Content Tax rates, compound interest, cost of living.

Activity The NBA and the NBA Players Union negotiated the value for the maximum contract that teams are permitted to offer free agents. The value for the maximum contract changes each year, and in 2016 the maximum value was $26,540,100. The most desirable free agent during the summer of 2016 was Kevin Durant, who interviewed with teams from six cities: Boston Celtics (BOS), Golden State Warriors (GSW), Los Angeles Clippers (LAC), Miami Heat (MIA), Oklahoma City Thunder (OKC), and San Antonio Spurs (SAS). It is fair to assume that each team offered Kevin Durant the maximum value, so he was left to choose a team based on factors other than gross pay. He eventually chose to sign a 2 year contract with the Golden State Warriors. Use the table below to answer the following questions, noting we will only be concerned with Durant's 2016 salary for this activity.

	BOS	GSW	LAC	MIA	OKC	SAS
Federal income tax %	39.7	39.7	39.7	39.7	39.7	39.7
State income tax %	5.10	13.3	13.3	0	5.25	0
Cost-of-living index	89.37	97	69.41	84.24	69.74	73.73

1. What is the net worth of the 2016 contract after paying federal and state taxes for each of the six teams? Assuming a biweekly paycheck schedule, what would each paycheck be worth?

2. On July 1, 2016, national interest rates for a savings account were extremely low, hovering around 0.03%. Even at that tiny rate, how much money could be made if the additional money from each paycheck relative to the GSW paycheck was deposited in a savings account that gave compound interest? Assume that the money is equally divided into 12 deposits, deposited on the first of every month, and that the interest is compounded monthly.

3. The cost-of-living index (CLI) measures how expensive it is to live in a specific city. It is most often used as a comparison to see what one would need to earn in a "new" city to afford the same goods as in an "old" city. The relative-change formula utilizing the CLI is

$$\frac{\text{New City CLI} - \text{Old City CLI}}{\text{Old City CLI}}.$$

This formula yields a percentage value representing how much more or less income you would need to earn relative to your income in your old city. Using this formula, calculate the necessary percentage increase or decrease in net income necessary to make playing for the five other teams equal in terms of relative-net income to playing for the Golden State Warriors.

4. Turn the above percentages into dollar amounts by multiplying each percentage by the net worth of his contract playing for the Golden State Warriors. If net worth from the 2016 contract was the only factor in deciding where to play, with which team should he have chosen to sign?

5. Obviously, other factors besides salary are considered when a person decides where to live and work. What other factors might have affected his decision?

5.4 Projects

5.4.1 Randon I. Grant

R.I.G. Agency
Randon I. Grant, chief agent
Hartford, CT 06103

Mathematical Thinking Student,

You likely have never heard of me, but I am the chief sports agent at R.I.G. Agency in Hartford, CT. I have built my agency negotiating the best contracts possible for sports stars in the upcoming Olympic sports of Pokemon Go and Coffee Pouring and now have landed my first mainstream sports star in the NBA point guard, Jeremy "Little Hands" Bole. Upon hiring me, Jeremy let it be known that he valued his talent to be worth on average $7,000,000 per year and wanted a 3 year contract worth a total of $21 million. Drawing upon my vast experience of negotiating contracts, I was able to secure contracts with NBA teams in five different cities each totaling $24 million.

Very proud of my work creating not only additional money but several options for my client, I presented all five contracts to him. Unfortunately, he was unhappy that I did not secure just one contract and asked me to tell him which offer was the "best" one for him. I was flabbergasted as I have no idea how to judge what is best for someone else, so I promised to try to help. And here is where I need your expertise: You see, I am great at negotiating with high-powered executives over a round of golf, but I have no idea how to actually compare the relative values of multiple offers. For obvious reasons I am not allowed to show you the specific contracts from each team; however, the table below shows you the yearly salary (in millions) I was able to secure from the team in each city.

	Brooklyn	Denver	Washington, DC	Milwaukee	Houston
Year 1	10	15	3	8	5
Year 2	10	4	7	8	9
Year 3	4	5	14	8	10
Total	24	24	24	24	24

As previously stated, the total salary came out to $24,000,000 for each offer, but the distribution of when the money is paid is different. What I need from you is a detailed report concerning the financial benefits of each offer, including but not limited to net pay after federal and state taxes, expected return on investment of all additional money in the five cities above the minimum net amount for each year, a comparison using the cost-of-living index, and the estimated mean and median wealth and size of the fanbase. In addition, as Jeremy will be moving to the new city with his family, including five children, we need some analysis of the top public and private schools for the areas in and around each city. Two of the children, Jeremy Jr. and Jerema, are beginning to look at colleges, so consider the in-state tuitions for the state schools and available scholarships at these institutions.

Thank you so much for your help, and I look forward to presenting your analysis to Jeremy myself.

Sincerely,
R. I. Grant
R.I.G. Agency

Implementation Notes The hardest part of this project for students was the comparative nature of the analysis. They have a difficult time understanding, after Year 1, the need to calculate the net pay for each city's offer relative to the salary offered by the Washington, DC team, and to use this relative value for the Year 1 investments. Some classes may benefit from a faculty member explicitly explaining why using Washington, DC as a floor for comparison for Year 1, Denver as a floor for Year 2, and Brooklyn as a floor for Year 3 is important.

5.4.2 Lucille Duck

Lucille Duck
San Antonio, TX 78278

Mathematical Thinking Student,

My name is Lucille Duck and I am an avid lottery player in my home state of Texas. Nicknamed "Lucky Ducky," I lived up to that calling by winning a super jackpot and now am looking for ways to better the "WOW" appeal of my home state. We like things big here, so I was was looking into building the world's largest "something" to have as a tourist attraction at the center of one of our great cities. While researching what this great something should be, I stumbled across the designs for the Tokyo Olympic Stadium.

The originally proposed Olympic Stadium for the 2020 Tokyo Olympics was estimated to cost $2.1 billion before being scrapped for a more economical $1.26 billion stadium. However, the original stadium was inspiring beyond belief and I simply can't live in a world without such beauty. Thus I have decided to pledge $500 million of my own personal wealth to help my favorite architect, Ms. Steel Smith, build the stadium. She feels that the ideal spot for the stadium would be San Antonio as the city is hoping to host a Summer Olympics in the near future and is currently trying to lure an NFL team to the city, so our Olympic Stadium could then be used for such future events. Now $500 million is far short of the necessary funds to build a $2.1 billion stadium, but at least it's a start.

I am asking you for an analysis of the best way to turn my initial investment into the full funds for building such a stadium within 10 years. Could you research at least three different ways other cities have raised money for stadiums or similar projects, such as selling bonds, product taxes (Philadelphia's soda tax, passed in 2016, would be a good example), or a new lottery, and provide expected revenues from each? In addition, how should my initial gift of $500 million be invested to generate the most wealth within 10 years? You should research national interest rates to see if a compound interest savings account would be useful compared with investing the money in an indexed stock like the S&P 500. You may also want to look at and discuss the recent 1 year, 3 year, and 5 year returns for the S&P 500 and if any world events might have affected the potential growth over those time periods.

Now, of course, building projects of this magnitude run into unexpected costs. The current mayor of San Antonio has agreed to finance all overrun costs above the $2.1 billion with a city sales tax increase, which would need special approval by the state senate. This tax would obviously be very unpleasant for the city, so I need you to research how much we should expect the cost of the stadium will run over budget assuming similar percentage overruns to the last few Summer Olympics. Then determine what percentage increase in the city sales tax would be needed to cover this cost over 5, 10, 20, or 30 years. Of course, if my gift grows sufficiently quickly, this may not be an issue, but keep in mind that an initial payment of all earmarked money must be made 10 years from now, so all investments will be cashed out at that point.

Good luck, and don't forget about looking into federal taxes on all generated money from investments and interest.

Sincerely,
Lucille "Lucky Ducky" Duck

Implementation Notes A brief review of percentages either in a traditional format or through activity in Section 5.3.1 will help set the stage for the last part of this project concerning cost overruns and city sales taxes.

5.4.3 Mone Y. Banks

Mone Y. Banks
Sports Finance and Investments
Tucson, AZ 85719

Dear Mathematical Thinking Student,

My company, Sports Finance and Investments, based in Tucson, AZ, has a meeting scheduled to present to the International Olympic Committee a proposal that outlines sound international financial planning for Olympic athletes. As countries reward their Olympians in myriad ways, we have had to spend a considerable amount of time on the research phase of our proposal, and we are running out of time as our meeting with the IOC in Switzerland is just weeks away. There are some countries that always send a significant number of athletes to the Olympics that we will not have time to research, so we would like to hire you to research the financial benefits Olympians from the United Kingdom, Canada, Germany, and Japan receive for representing their respective countries. A few items to keep in mind are listed below.

- What are the Olympic medal bonuses from each of the national Olympic Committees, standardized to the common currency of euros?

- Assuming gross pay for bonuses is taxed as income by each country and, if applicable, by the state or province, what is the net pay for each bonus in the four countries? For state or province tax rates, assume that each athlete lives in the capital city of their country.
- How does this income compare with the mean and median incomes of citizens of each country?
- Are there other sources of income from the government, such as training stipends, that should be considered?

Additionally, occasionally athletes from different countries then decide to form families. Suppose that a British and a German athlete are married, as are a Canadian and a Japanese athlete. Where should each couple invest their winnings, assuming each athlete has a savings account and stock investments in her or his home country? Which countries offer the highest national interest rates on savings accounts and CDs, the lowest inflation rates, or the lowest national income tax rates? What about each country's stock market: do they grow at comparable rates and have comparable volatility? For each country, pick an index fund on their stock exchange and determine how much an investment would be worth today if a gold-medal-winning athlete from that country had put all of her or his winnings into that index fund 5, 10, 20, or 30 years ago.

Thank you in advance for lending us your research and presentation skills as we prepare for our meeting with the IOC.

Sincerely,

Mone Y. Banks

Implementation Notes When researching how much money each country pays its athletes for winning different medals, it is important to remind students to try to find the information all for the same year. As many of the calculations are repeated for each country, emphasizing the means of presentation of data can help students focus on drawing comparisons between the countries.

6 Data

In this chapter, we present five activities and three projects that are concerned with the representation of data, summary statistics such as measures of central tendency and spread, and distributions. The sports chosen to accompany this material are track and field, baseball, and golf.

6.1 Mathematics

Representations of Data

- Frequency representations—histogram, line graph, bar graph
- Numerically conserving representations—Bar graph, line graph, dot plot, stem and leaf plot
- Proportional representations—pie chart, pictograph

Summary Statistics

- Center—mean, median, mode
- Five-number summary—interquartile range, box plot

Distributions

- Describing variability—standard deviation
- Distributions—normal and skewed curves, bimodal and unimodal curves

6.2 Sports

Track and Field Two activities in this chapter will be concerned with running, particularly the Boston Marathon and the women's 1500 meter race, while one project in this chapter will be concerned with the 100 meter dash and the marathon race. The 100 meter dash is a sprint where, from a stationary position, competitors run on a straight track the distance of 100 meters. The unofficial title of the world's fastest man or woman is given to the winner of the 100 meter dash performed at the Summer Olympics. The race was part of the first modern-era Olympics, held in Athens in 1896, and was won by the American sprinter Thomas Burke.[1] The 1500 meter race is considered a medium-length race. This race was also part of the Olympics of 1896 in Athens and was the final event on April 7. Trailing for most of the race, the Australian Edwin Flack overcame the leader in the last 100 meters to win with a time of 4:33.2 minutes, a mere 5 meters and 0.4 seconds ahead of the American Arthur Blake.[2] The marathon is a long-distance race, usually run through a city or across towns and countryside. The race was also a part of the 1896 Olympics, where the race was based on the legend of Philippides from ancient Greece and was run from Marathon Bridge to the Olympic Stadium, with the total length being 24.85 miles. At the 1908 Olympics in London, the race was extended to 26.33 miles so that the conclusion was in front of the royal family's viewing box, and, by the 1924 games in Paris, the official marathon distance of 26.219 miles was established.[3]

Baseball One activity and one project in this chapter will be concerned with the hitting statistics of on-base percentage, slugging percentage, and runs batted in. *On-base percentage*, often denoted OBP, is the ratio of the number of times a hitter reaches base to his plate appearances. This statistic treats all hits equally and so does not disadvantage hitters that do not try to hit home runs and who may score a run when the next batter gets a hit. The all-time leader in career OBP is Ted Williams, with an OBP of 0.482, while the single-season record is held by Barry Bonds, with an OBP of 0.609 in 2004.[4] In contrast, *slugging percentage*, often denoted SLG, is a weighted average of a batter's hits and is defined to be the total number of bases divided by the number of at bats. The fact that this statistic places more emphasis on extra-base hits which more directly produce runs for the batting team makes it a popular measure of a hitter's value. The all-time leader in career slugging percentage is Babe Ruth, with an SLG of 0.690, while the single-season record is held by

[1] SR/Olympic Sports. "Athletics at the 1896 Athina Summer Games: Men's 100 metres." https://www.sports-reference.com/olympics/summer/1896/ATH/mens-100-metres.html (accessed: December 3, 2018).
[2] SR/Olympic Sports. "Athletics at the 1896 Athina Summer Games: Men's 1,500 metres." https://www.sports-reference.com/olympics/summer/1896/ATH/mens-1500-metres.html (accessed: December 3, 2018).
[3] Exercise the Right to Read. "The History of the Marathon." http://www.exercisetherighttoread.org/historyofmarathon.pdf (accessed: July 8, 2014).
[4] Baseball Reference. "Career Leaders & Records for On-Base %." https://www.baseball-reference.com/leaders/onbase_perc_career.shtml (accessed: December 3, 2018).

Barry Bonds, with an SLG of 0.864 in 2004.[5] On the other hand, *runs batted in*, or RBIs, counts the number of times a hitter at bat directly results in a run being scored while no fielding errors occurred and fewer than two outs were recorded on the play. This makes RBIs a statistic that measures an absolute quantity rather than an average quantity. Hank Aaron holds the record for the most RBIs in a career at 2,297, while the single-season record of 191 RBIs was accomplished in 1930 by Hack Wilson.[6]

Golf Two activities and one project in this chapter are concerned with the Masters tournament in golf. The Masters is one of the most prestigious golf tournaments in the world and is played in Augusta, Georgia, usually in early April. Winners of the tournament receive a golf medal, a sterling replica of the Masters trophy, and, most recognizably, a green jacket. The tournament was first played in 1934 and was won by Horton Smith, who also won the tournament in 1936. The player who has won the most Masters tournaments is Jack Nicklaus (6), and only Nicklaus (1965–66), Nick Faldo (1989–90), and Tiger Woods (2001–02) have won consecutive Masters tournaments. Many policies concerning tournament golf have their origins at the Masters, including playing 18 holes on four consecutive days.[7] Each hole of a golf course is rated by the expected number of strokes, both full swings and putts, necessary for a professional golfer to hit the ball into the cup; this score is called the *par*. A *putt* is a golf stroke made on the area surrounding the hole, commonly called the *green*, that causes the ball to roll into or around the hole.[8] A score is called a *birdie*, first termed thus in 1903 at the Country Club in Atlantic City, if a golfer uses one fewer stroke than par to hit the ball into the cup.[9]

6.3 Activities

6.3.1 Men's Boston Marathon

Content Five number summary, box plot, measures of center.

Activity In 1980 Bill Rodgers won his fourth Boston Marathon, with a time of 2 hours, 12 minutes, and 11 seconds (written 2:12:11); only Robert Cheruiyot of Kenya (five, last in 2010), Gerard Cote of Canada (four, last in 1948), and Clarence DeMar of the

[5] Baseball Reference. "Career Leaders & Records for Slugging %." https://www.baseball-reference.com/leaders/slugging_perc_career.shtml (accessed: December 3, 2018).
[6] Baseball Reference. "Career Leaders & Records for Runs Batted In." https://www.baseball-reference.com/leaders/RBI_career.shtml (accessed: December 3, 2018).
[7] Masters. "Historical Records and Stats." http://www.masters.com/en_US/scores/stats/historical/index.html (accessed: December 3, 2018).
[8] Merriam-Webster.com. "Putt." 2014. http://www.merriam-webster.com/dictionary/putt (accessed: July 15, 2014).
[9] Scottish Golf History. "Bogey to Blow-Up." http://www.scottishgolfhistory.org/origin-of-golf-terms/bogey/ (accessed: July 15, 2014).

United States (seven, last in 1930) have as many.[10] In this activity you will analyze the winning times from Bill Rodgers' 1980 race until 2014, which is the last year that an American, Meb Keflezighi, has won the race since 1983. Over this time span the course record for fastest time has been broken seven times, and one of those times, Geoffrey Mutai's 2:03:02, was the world record. Table A.5[11] displays each champion's race time and can be found in the "Data Sets" appendix.

1. Create a five-number summary for the champion's race times from 1980–2014. Are all aspects of the racer's time relevant to the five number summary?

2. Create a box plot with an appropriately scaled axis using data from Question 1.

3. Using the interquartile range, are there any outliers among the champions' times? What could have been possible causes for the particularly slow or fast times among the world's top marathoners?

4. Calculate the center of the champions' times. Which of the measures (mean, median, and mode) used to determine the center is the least useful, which is the most useful, and why?

5. What information from the chart could be analyzed in a meaningful manner using the measure deemed the "least useful"?

6.3.2 Woman's Olympic 1500 Meter Race

Content Frequency tables, histograms, bar graphs, pictographs.

Activity The women's 1500 meter race has been run in the Olympics from 1972 to the present day. The Olympic record for this event was set in 1988 by the Romanian middle-distance runner Paula Ivan, who won gold that year with a time of 3:53.96 minutes. The event has three stages: a heats stage, a semifinals stage, and a finals stage. At each stage, all women left in the competition run the race, with those best times advancing to the next stage while the others are all eliminated. The finals stage typically is run by 12 women, where the woman with the fastest time wins the gold medal, the woman with the second fastest time wins the silver medal, and the third fastest runner wins the bronze medal. In this activity, you will develop multiple representations for data by analyzing the countries of origin of the medalists in the Olympic women's 1500 meter race. A table[12]

---

[10] Boston Athletic Association. "Champions of the Boston Marathon: Men's Open Division." https://www.baa.org/races/boston-marathon/results/champions (accessed: December 3, 2018).

[11] Ibid.

[12] International Olympic Committee. "1500m Women." https://www.olympic.org/athletics/1500m-women (accessed: December 3, 2018).

containing the countries of origin of the medalists for this event from 1972 through 2016 is below.

Nationality	Abbreviation	Gold	Silver	Bronze
Algeria	ALG	2	0	0
Austria	AUT	0	0	1
Bahrain	BRN	1	0	0
China	CHN	0	0	1
East Germany	GDR	0	3	1
Ethiopia	ETH	0	1	1
Great Britain	GBR	1	0	0
Italy	ITA	1	0	1
Kenya	KEN	2	0	0
Romania	ROU	1	3	3
Russia	RUS	1	2	0
Soviet Union	URS	3	1	2
Ukraine	UKR	0	1	1
Unified Team	EUN	0	1	0
United States	USA	0	0	1

1. The table above is an example of a *frequency table*, where one can easily determine the frequency of an event in a data set. In this case, you can see the frequency with which a country earned a gold, a silver, or a bronze medal in the women's 1500 m race. Create your own frequency table of two columns to display the total number of medals each country has earned in this race.

2. Frequency tables can naturally be turned into representations: Create three separate copies of two perpendicular axes. Label the horizontal axis with the countries given in the table above and label the vertical axis with a scale appropriate to the data. Put the headings "Countries" and "Frequency" on the appropriate axis. Now how might one try to create a graphical representation of your "Total medal" frequency table? What would be an appropriate scale for the vertical axis? Brainstorm some ideas but don't put anything to paper yet; just discuss within your group.

3. Now you're going to create three different representations, one on each axis.

   (a) On the first pair of axes, draw a ★ above each country for each medal it earned, placing the ★'s in a single-file column. Draw them appropriately sized so that they align with your vertical frequency axis.

(b) On the second pair of axes, draw a single ⋆ above each country at the height of the total number of medals it earned. Then connect adjacent ⋆'s with straight edges.

(c) On the third pair of axes, draw a rectangle above each country to the height of the total number of medals it earned. The rectangles should all be equal in width, and there should be gaps between them.

4. The three representations you have created are called a pictograph, a line graph, and a bar graph, respectively. What are some of the similarities and differences between these representations? There is a characteristic of the line graph that is slightly misleading for this data and makes it a less than ideal choice for representing this type of data; what is it?

5. Obviously, these representations hide some aspects of the original data set, particularly which medal was won. However, each can be altered to include this information.

(a) Create two separate copies of two perpendicular axes using the same labels and scales as before.

(b) On the first one, we're going to alter the pictograph representation. Instead of having a single column of ⋆'s for each country's medal count, above each country's label create three columns: one with ∗'s corresponding to the number of gold medals, one with o's corresponding to the number of silver medals, and one with ◇'s corresponding to the number of bronze medals. This new graph is still an example of a pictograph.

(c) On the second pair of axes, alter your bar graph similarly by creating three equal-width rectangles above each country, the left one to correspond to gold medals, the middle to silver medals, and the right to bronze. It will be helpful to differentiate the shades or colors for the left, middle, and right columns for each country. The three columns above a single country should all still be touching but now have small gaps between the columns for different countries. This new bar graph represents multiple data sets (gold, silver, and bronze medals).

6. If you had to rank the top three countries by their athletes' performances in this race, how would your different representations lead you to different rankings? Create four different rankings: one that takes into account only gold medals, one that takes into account gold and silver medals, one that takes into account gold, silver, and bronze medals, and one that takes into account only the total number of medals but does not differentiate in any way which medal was earned. What arguments can be made to support each type of ranking?

6.3.3 2013 Red Sox Hitting

Content Mean, weighted averages, numerically conserving representations.

Activity For this activity we will be looking at some hitting statistics compiled from data for the 2013 Red Sox team that won the World Series. This team led the American League in many important hitting statistics, including runs, runs batted in, doubles, on-base percentage, and slugging percentage, while finishing second or third in hits, triples, stolen bases, walks, and batting average.[13] In this activity we will restrict ourselves to batters who accumulated 300 or more at bats during the season, and a table containing the hitting statistics for these 2013 Red Sox players can be found in the "Data Sets" appendix as Table A.6.[14]

1. Between batting average, OBP, and SLG, do you expect one statistic to always be the lowest value for each player? What about one that should be consistently the highest? Why do you expect these results?

2. Calculate each player's batting average, OBP, and SLG.

3. Was your intuition correct? Discuss any counterexamples that may have occurred.

4. Another statistic that is commonly used to evaluate batters is called *on-base plus slugging* and is denoted OPS. This statistic is simply the sum of a player's OBP and SLG.[15] Calculate each 2013 Red Sox player's OPS from your answer to Question 2.

5. Would a dot plot, stem and leaf plot, or bar graph be the most effective way to show OBP, SLG, and OPS for each player in a single representation? Create a representation that shows all 27 data values.

Implementation Notes As this activity is computationally heavy, a calculator or computer is highly recommended.

6.3.4 Masters Tournament Champions

Content Pictograph, length-versus-area relation.

Activity A pictograph is a type of representation that uses a picture or diagram to convey information. In everyday life, we interpret pictures such as traffic signs, computer icons, and safety warnings to understand very specific and sometimes complex messages in a time-efficient manner. It is important that everyone who sees these types of picture understands the same message, so that information can be communicated even between people who speak different languages. In order to represent a distribution, pictographs may either repeat symbols a specified number of times or increase the size of the picture to show higher numerical values.

[13] Baseball Reference. "2013 AL Team Statistics." http://www.baseball-reference.com/leagues/AL/2013.shtml (accessed: July 23, 2014).
[14] ESPN. "Boston Red Sox Batting Stats—2013." http://www.espn.com/mlb/team/stats/batting/_/name/bos/year/2013 (accessed: December 3, 2018).
[15] Baseball Reference. "OPS." http://www.baseball-reference.com/bullpen/OPS (accessed: July 23, 2014).

For this activity we will be concerned with the champions of the Masters tournament. In total, 48 different golfers from 11 countries have won the 78 Masters tournaments between 1934 and 2014. (The tournament was not played in 1943–1945 due to World War II.) The table[16] below organizes this information in a concise manner.

Nationality	Number of wins	Number of winners
United States	58	35
South Africa	5	3
Spain	4	2
England	3	1
Germany	2	1
Scotland	1	1
Wales	1	1
Fiji	1	1
Canada	1	1
Argentina	1	1
Australia	1	1

1. Create a new three-column chart similar to the table above but, instead of using digits to represent the number of wins and number of winners for each country, use stick figures. For which countries does your chart offer an easy-to-read representation and for which is it less useful?

2. To try to make your chart more useful, you may create a key where each stick figure represents multiple wins or winners. Recreate the chart so that each stick figure represents two wins or winners; you will need to use a "half stick figure" to represent odd values. Does your key make the chart easier or harder to interpret? How did you divide your stick figure to represent a single winner? Would dividing the picture vertically or horizontally give a better representation of half the key value?

3. Instead of increasing the number of stick figures, another option to represent relative size is to represent larger numerical values by using larger figures. Make a key where a small rectangular box, a flag, represents one win or winner. A natural diagram can be created for each country by drawing a square box with the country's initial(s) inside, where the length and width of the box both stretch. For example, in the number-of-wins chart, Fiji should have a 1×1 box while Spain should have a 4×4 box. After you finish, what strikes you as incorrect about this representation? How many more wins does it appear Spain has over Fiji?

16 Wikipedia. "List of Masters Tournament Champions." https://en.wikipedia.org/wiki/List_of_Masters_Tournament_champions (accessed: July 15, 2014).

4. When you look at the flags in your chart, do you gain a sense of size from one dimension or two dimensions? Recreate your chart with flags so that the relative size of each flag accurately represents the number of wins or winners.

6.3.5 2014 Masters Tournament Scoring

Content Dot plots, pie charts, distribution shapes, standard deviation, inferences.

Activity The 2014 Masters golf tournament was played in ideal weather, with no rain and only light winds during the four-day tournament. As the environmental challenges of each hole were consistent from one golfer to the next, analysis of players' scores during this year may more accurately represent how difficult the course played than in years with more variable weather.

In this activity you will analyze two aspects of a golfer's round, the number of birdies and the number of putts, from the 2014 Masters tournament. Tables A.7 and A.8 display the total numbers of putts and birdies per hole for each golfer and can be found in the "Data Sets" appendix.

1. Create two separate dot plots, one comparing each golfer's total number of birdies and one comparing each golfer's putts per hole average at the 2014 Masters.

2. Impose a smooth-curve approximation on your dot plots from Question 1. Are the curves unimodal or bimodal, or do they have some other shape? Why might the distribution of putts per hole have a different shape than the distribution of total birdies?

3. Determine the distribution of the number of birdies per 18 holes at the 2014 Masters; it may be helpful to create a dot plot representation of it. In paragraph form, discuss which of the previous distributions you expect this new distribution to be most similar to in terms of shape.

4. After the second round of the tournament, only the top percentage of players are allowed to play in the final two rounds. Using only the data from Rounds 1 and 2, determine the standard deviation of the number of birdies scored by each player and the standard deviation of putts per 18 holes for each player. Does it appear that either of these standard deviations could be used to indicate who was allowed to play in the last two rounds? What are some possible explanations for any exceptions?

5. Using only the data for the players who were allowed to play in all four rounds of the tournament, create pie charts to represent this data. One pie chart should show in which round each player had the most birdies and the other should show in which round each player had his lowest putts per hole. (Choose the highest-numbered round if there is a tie.) Does there seem to be a round that played the easiest and one that played the hardest? Knowing that weather was not a factor during this tournament, what are some possible explanations if there did appear to be a hardest or easiest round?

Implementation Notes A calculator or computer may be useful as an aid for this activity.

6.4 Projects

6.4.1 S. Peter

Happiness Creamery
S. Peter, Manager
Bloomsburg, PA 17815

Introduction to Statistics Student,

I'm Steven Peter, Spetey for short, and I own the soon-to-be world-famous Happiness Creamery. Our business provides home delivery service of everyone's favorite summer dessert, ice cream. The idea is simple: after working all day outside in the midsummer heat, nobody really wants to walk inside and order a hot pizza that will only increase their discomfort. Instead, folks do wish to be able to order a frozen treat and have it delivered in 45 minutes or less. To be even more convenient for the customer, we situated our business inside the town park so picnickers, frisbee golfers, and softball players can call us in the park and have one of our delivery people literally run the ice cream to them. Such a simple and brilliant idea, I know it's hard to imagine that nobody thought of it first.

However, last week we stumbled upon a problem. An order was placed for one of our running delivery services to a house that was 1 mile away from our location. We have two ice cream runners: Quail Wic, a 100 meter dash runner in college who is used for the short park deliveries, and Paula Ace, who ran marathons and is used for the deliveries downtown, which is 3–4 miles up the road. When an order comes in that is not in either Q. Wic or P. Ace's normal delivery distance, their discussion concerning who should deliver the ice cream sometimes digresses into a discussion concerning running, and this is what happened:

Wic claimed that, since the 100 m dash has been used to unofficially determine the fastest woman in the world and, over the last 25 years, the countries of Jamaica and the United States have consistently trained the fastest woman, then this distance most accurately measures dependability, and therefore she should get to deliver the order. Ace, who happens to be from Kenya, immediately took offense, noting that while the woman with the fastest 100 m time often wins that title four or five years in a row, the fastest marathon runners rarely hold that title for more than two consecutive years. Thus a country's consistently strong performance is more indicative of quality training rather than of a single superb athlete, and so she should get to deliver the ice cream. Unfortunately, by the time the two ladies decided to flip a coin to see who won the delivery, my wonderful banana split sundae had melted into a puddle of milk and chocolate fudge. The argument has continued all week and more of my frozen treats are succumbing to the heat of summer.

What I would like from you is support for both of their arguments. First, an analysis of the best yearly women's 100 m dash times from 1980. Since we are most interested in the country of origin of the fastest woman for each year, it would be nice to see both a pie chart and a pictograph using countries' flags as icons to represent the data. In many of these years the Summer Olympics were held, including 1980, 1984, 1988,..., so please include a discussion of how often the country that won the gold medal in the 100 m dash also had the fastest woman that year. This will give credence to Wic's argument. Next, to support Ace, repeat your analysis for the women's marathon. Does one distance seem to be more difficult for a country to win the yearly record and the Olympic record?

I thank you for your help, and please hurry, the ice cream is melting!

Sincerely,
Steven Peter
Happiness Creamery

6.4.2 R. B. Insley

R.B.I. Hitters Academy
Ruby Brandon Insley, Founder & Owner
Hingham, MA 02043

Mathematical Thinking Student,

I am R. B. Insley and am the founder of the world-famous baseball training facility "R.B.I. Hitters Academy" in Massachusetts. Here at R.B.I., I have been training 8–17-year-old girls and boys how to properly approach hitting in baseball since 1981. After the hitting boom seen in Major League Baseball (MLB) in the 1980s and early 1990s, my company grew in scope and size to the point where we started concerning ourselves with not only hitting for power but also bunting, running the bases, stealing signs, and, most impressively, scouting. Although welcomed, this growth meant that I could no longer run the company by myself, and I hired four other people, one for each new department, to join me on a board of directors to run the company.

Just last week, my company's vice president of advanced statistics, Dr. Beam, suggested to the board that we disregard the art of hitting for power and refocus the hitting aspect of the academy solely on sacrifice bunting. I am sure this suggestion is a means to minimize my skills as a hitting champion, eventually leading to my dismissal from the governing board, and I can't let this happen! The board meets in a few weeks to discuss this suggestion. I'll be allowed to make a counterargument before the official vote occurs. I am sure that Dr. Beam has convinced the entire advanced statistics department to sabotage my presentation, so I need you and your expertise in summary statistics and distributions to help me prepare my presentation.

First, you will look at the RBI leaders for MLB since 1900. I need you to create a chart that denotes the maximum RBIs a player created in each year from 1900 through 2014. Then, for the numerically inclined members of our board, include the center of this data set and the standard deviation of the data set consisting of the 115 year-maximums. To aid the visually included members, create a histogram, a line graph, and a smooth-curve approximation of the distribution.

Second, I need you to repeat this process but, instead of taking the maximum number of RBIs for each year to create your data set, use the average of the maximums for each decade and repeat your analysis on this new distribution of 12 data points. For example, the maximum RBIs for the years 1900–1909 were. 110, 126, 121, 104, 102, 121, 96,

119, 109, 107, so the average of these values which would be used for this data set is 111.5.

I recognize your dedication to R.B.I. Hitters Academy and await your report.

Sincerely,
Ruby Brandon Insley (R.B.I)
R.B.I. Hitters Academy

6.4.3 Ray Ling

Exterior Designs and You
Ray Ling, Wrought Iron Specialist
Windermere, FL 34786

Bloomsburg University Student,

As you may know, the golf sensation Bubba Watson lives in Windermere, FL. The town originally commissioned me to make a statue celebrating his second time winning the Master's golf championship, which occurred on April 13, 2014. Unfortunately, one of our fiscally minded residents protested the use of public funds for the statue at our town hall meeting. Since this meeting, I have been asked to produce a report that describes how impressive Bubba's second Masters championship was in historical context. I would hate to see all of my hard work wasted, but the pressure of producing this report while continuing to work on the statue is just too much for me to handle at one time.

I am close friends with your instructor Dr. Woods and originally requested he work on producing this report for me. However, he declined due to a coming sabbatical, where he plans to study the correlation between increased golf activity and blood pressure medicine. However, he suggested you as a willing substitute, I heard that you have studied distributions and variability in his class and should be well prepared to write the report.

Thus I would like you to work out the details of the argument I can use to support this public works project. Since the report needs to compare Bubba's championship with previous championships, I think the best idea would be to look at the scores relative to par. In particular, does the score relative to par of the champions from 1946 to the present day have a normal distribution, and should the score of −8 by Bubba Watson to win the 2014 tournament be considered an average, above-average, or below-average champion's score? Please provide a smooth-curve approximation of the distribution of the champions' scores along with the standard deviation and where Bubba's score falls. Also, Bubba Watson was in second place after the first round but then was either in first place or tied for first after each of the final three rounds of the tournament. Determine if Bubba's comeback from second place to win the tournament after the first round is an impressive feat by looking at the position after the first round of each of the eventual champions from 1946 to the present day and considering the smooth-curve distribution of the champions's first-round placement.

I appreciate your help in this matter.

Sincerely,
Ray Ling

7 Statistical Reasoning

This chapter contains five activities and three projects addressing introductory statistical design, including sampling techniques, correlation versus causation, and basic statistical inference. The content is approached through the sports of swimming, tennis, and baseball.

7.1 Mathematics

Correlation and Causation

- Scatter plots
- Trends–positive, negative, weak, strong, linear, nonlinear

Introductory Statistical Design

- Sampling—opportunity, systematic, simple random, stratified
- Randomness

Inference

- Margin of error and confidence intervals
- Hypothesis testing

7.2 Sports

Baseball Major League Baseball statisticians, amateur and expert alike, are known for tracking a wide range of statistics. The use of sabermetrics has only increased since

Oakland A's manager Billy Beane effectively used modern individual-player statistics such as on-base percentage to build a low-budget team which could compete with teams spending much more.[1] In this chapter, an activity and a project relate to statistics in baseball. First, a pitcher pitches a *complete game* when he pitches the entire game without being relieved by another player. The activity will deal with the frequency of complete games in MLB. In the project, students will compare the batting statistics of hits per game (H/G), walks per game (BB/G), and strikeouts per game (SO/G) of today's baseball teams with the average statistics from 1915.

Tennis Two activities and one project will deal with data from major tennis tournaments. Professional tennis is governed by two groups: the Association of Tennis Professionals (ATP) for the men, and the Women's Tennis Association (WTA) for the women. Every year, these associations organize four major events known as Grand Slam tournaments. These tournaments offer significantly higher prestige, viewership, and purses than any others held throughout the year. Thus, these tournaments attract the best players in the world. The four tournaments are the Australian Open, the French Open, Wimbledon, and the US Open. Each of these tournaments has a starting field of 128 men and 128 women, competing for men's and women's singles titles. Of course, this leads to 127 men's matches and 127 women's matches to determine the winners. The two activities in this chapter will concentrate on the French Open and Wimbledon, and the project will compare wins in the Australian Open and US Open.

Swimming Four strokes have been recognized in the Summer Olympics since 1968. These are freestyle, backstroke, breaststroke, and butterfly. Men and women compete individually and in relays to produce the fastest times while swimming varying distances and using different strokes or combinations of strokes. In the activity, we consider the 4×200m freestyle relay. For this event, a team of four men or four women swim a relay where each leg is 200 meters long. All competitors swim the freestyle stroke. The project is concerned with swimming events in the Olympics and will focus on comparing individual finish times among the different strokes.

7.3 Activities

7.3.1 MLB Complete Games in 2013

Content Sampling methods: convenience, systematic, simple random, stratified.

Activity In early baseball history, complete games were quite commonplace; however, today, with a more high-powered game and fear of injuries, most MLB pitchers are relieved before the end of the game even if they are pitching well. In this project, we

---

[1] Lewis, Michael. *Moneyball: The Art of Winning an Unfair Game.* New York: W.W. Norton 2004.

use different sampling methods to estimate the number of complete games in the 2013 season, which was 184 days long. Complete-game data for the 2013 season can be found in Table A.9.

1. In this study we will use the number of complete games pitched over a subset of 14 days during the season to predict the total number of complete games for the season. What is the population and what is the sample in this study?

2. (a) How many complete games were pitched in the first 14 days? Use this number to approximate the total number of complete games for the entire season.

   (b) The sample consisting of the first 14 days is a *convenience sample* because these days were chosen from the population by selecting days that were easily available or convenient. What is another way you could choose a convenience sample of size 14 from the 184 day season?

3. (a) How many complete games were pitched every other Monday, beginning with the first Monday of the season? Use this number to approximate the total number of complete games for the entire season.

   (b) Choosing every other Monday is called a *systematic sample* because we are systematically choosing every *k*th member of a population. What is another way you could choose a systematic sample of 14 days from the 184 day season?

4. (a) Use a random number generator (such as the one found on random.org) to randomly choose 14 of the 184 days. How many complete games were pitched on those days? Use this number to approximate the total number of complete games for the entire season.

   (b) A random number generator chooses a *simple random sample* because every sample of the same size has an equally likely chance of being chosen. How else could you choose a simple random sample of size 14?

5. (a) Divide the season in half, from March through June and from July through September. Take a simple random sample of 7 days from each half and total the number of complete games in your two samples combined. Use this number to approximate the total number of complete games for the entire season.

   (b) We have taken a *stratified sample* by splitting the population into groups, taking a subsample from each group, and combining those samples into a bigger sample used to describe the entire population. What is another way to take a stratified sample of 14 days?

6. There were 127 complete games pitched in the 2013 season. Which sampling method (as listed in the table below) most accurately estimated the number of complete games? Discuss why each sampling method was or was not a good estimator for this data.

Sampling method	Good estimator (Y/N)	Why or why not?
Convenience sample		
Systematic sample		
Simple random sample		
Stratified sample		

7. A *representative sample* is a sample which will have the same relevant characteristics as the population. With this in mind, can you draw conclusions about the number of complete games in the 20th century (1901–2000) using one of the samples described in the chart above? Explain why a statistical study must use a representative sample in order to draw conclusions about a population.

7.3.2 Number of Games in a Match at Wimbledon

Content Randomness, margin of error, confidence intervals.

Activity Tennis matches are composed of individual points that can be won by either player. These points are combined into games, which in turn make up sets, and, finally, the match is made up of sets. The longest tennis match on record took place at Wimbledon in 2010 between John Isner and Nicholas Mahut, lasting 183 games and taking 11 hours and 5 minutes over the course of 3 days.[2] While not expecting this record to be broken, the organizers for this year's tournament want to know what proportion of the matches are likely to go over 40 games. Table A.10 shows the number of games in each of the 127 men's singles matches at Wimbledon in 2014.

1. Using a simple random sampling technique such as a random number table or generator, choose a sample of 25 matches from Table A.10. What proportion of matches in your sample have more than 40 games?

2. Find a 95% confidence interval for the true population parameter using the approximation $1/\sqrt{n}$ for the margin of error.

3. Pick another sample of 25 matches from Table A.10, this time *not* using any scientific sampling technique such as SRS or systematic sampling. What proportion of matches in this sample have more than 40 games?

4. Find the 95% confidence interval for your nonrandom sample proportion.

5. The actual proportion of matches over 40 games at Wimbledon in 2014 was 44.9%. Did either or both of your confidence intervals contain the true population proportion?

---

[2] Wikipedia. "Isner–Mahut match at the 2010 Wimbledon Championships." https://en.wikipedia.org/wiki/Isner--Mahut_match_at_the_2010_Wimbledon_Championships (accessed: December 3, 2018).

6. Write your confidence interval and sample proportion for each type of sample on the board. For the class data, what percentage of the simple random sample confidence intervals contained the true proportion? What percentage of the other sample confidence intervals contained the true proportion? Is this what you expected?

7. Is one sampling method more likely to capture the population parameter? Which one?

8. Explain why a random sampling technique must be used to draw strong conclusions in a statistical study.

Implementation Notes We hope to show that the random sampling method is more likely to provide an unbiased estimator of the population parameter, but, of course, sometimes the nonscientific method will provide the better estimator, especially in small classes. You will want to be prepared for such a discussion if that happens with your class.

7.3.3 Unforced Errors in Tennis

Content Confidence intervals, margin of error, hypothesis testing.

Activity Like many sports, the game of tennis has a wealth of statistics used to evaluate many aspects of the game. Examples of these are break points saved or won, aces per match, and service games won. In this activity we will look at unforced errors. An *unforced error* in tennis is when the player loses a point due solely to his or her poor execution or mistake in judgment. In comparison, a *forced error* is the loss of a point due to an opponent's aggressive play. This statistic, introduced in the 1980s, has a checkered past. Since it is a subjective statistic, some players feel that it can vary widely from match to match and thus does not have any value. Nevertheless, unforced errors are a popular statistic with fans and appear on most stat sheets or television graphics during a match.

When comparing unforced errors from player to player or match to match, the total number of errors at the end of the match may not be the most informative statistic. This fact becomes obvious when we consider; that matches are not of uniform length; that is, the numbers of sets and of games within each set that each pair of opponents play can vary. We therefore consider a new statistic called *unforced errors per hour*, or UEpH. During major tennis events such as the French Open, there is a 20 second clock between points, forcing a more uniform duration for each point. Because of this, our new metric should be able to more accurately compare errors even though some players play fewer games and sets than other players. In particular, we can compare men's and women's matches despite the fact that women need only to win two sets to win a match while men need to win three sets.

Table A.11 contains 254 data points from the 2014 French Open for each player's UEpH for each game in which he competed.

1. Take a representative sample of 25 players' UEpH from Table A.11. Describe your sampling method.

2. Find the sample mean and standard deviation of the sample.

3. Compute the margin of error using the estimate $2s/\sqrt{n}$ and give the 95% confidence interval for the true mean UEpH for the men's singles tournament.

4. If the average UEpH for a woman in the tournament is 16.9, do you expect this parameter to be significantly different for men? Explain.

5. How would you change your experiment if you wanted your margin of error to be smaller?

A random sample of 100 players' UEpH selected from the 2013 ATP world tour shows that the mean UEpH is 15.6, and the probability that mean is less than the WTA world tour mean of 16.9 is 0.0186.

6. Formulate the null and alternative hypotheses for this situation.

7. Discuss whether there is evidence to reject or fail to reject the null hypothesis.

Implementation Notes This activity was used in a quantitative skills course, but there are many opportunities to modify it for use in an introductory statistics course, including the use of a more exact margin of error and more student development of the hypothesis test used in the last two questions.

7.3.4 Swimming 4×200 Relays in the Olympics

Content Scatter plot, correlation, trends, causation.

Activity The 4×200 m freestyle relay consists of four swimmers each swimming 200 meters, one after another. The top 18 national relay teams for both the men and the women qualify to race in the Olympics. The first round consists of two heats of eight teams, in which four teams from each heat move on to compete in a final of eight teams. When watching the Olympics, you often notice men's and women's teams from the same nation standing on the podium, so we wonder: is there a correlation between men's and women's relay times?

1. Consider the data from the 2012 Olympics found in Table A.11: what variables will go on the horizontal and vertical axes? What is an appropriate scale?

2. How can you deal with the nations for which one of the relays did not qualify for the Olympics?

3. Create a scatter plot relating men's and women's 4×200 freestyle relay times.

4. Describe the trend or lack of trend. Are there any influential points?

5. Are the men's and women's times correlated? Explain.

6. Do men's times have a causal relationship with women's times? Explain.

7. How do you predict the scatter plot would change if you collected relay finish times for the past five Olympic Games? Explain.

8. How do you predict the scatter plot would change if you included the qualifying times for all nations fielding relay teams, including those that did not qualify for the Olympics? Explain.

Implementation Notes Additional questions could involve calculation of the correlation coefficient and regression line, as well as making predictions based on the line. Also, note that you may want to discuss Question 1 together, as a few students will persist in drawing two separate plots in order to include all the data points in each column of the table.

7.3.5 Swimming World Records

Content Descriptive statistics, graphical representations.

Activity The first world records in the men's and women's 100 m freestyle in long course were recorded and recognized by the International Swimming Federation in the early 1900s.[3]

1. Using your favorite search engine, look up and record the world record progression for the men's or women's 100 m freestyle in long course, that is, swimming 50 m laps.

2. How many times has the record been broken?

3. What is the average margin by which the record is broken?

4. What is the median number of years between events where a new world record is set?

5. Make a graph to illustrate the world record times from 1900 to today. Describe some interesting features of your graph.

6. Do you think swimmers will continue to break world records at approximately the same rate as shown in your graph? Explain.

Implementation Notes This activity does ask the students to complete some research in class, but only one source is needed and these world record times can be found easily with a smartphone or other internet-equipped device. The long-course data set contains around 50 data points. For shorter class periods, you may wish to modify the activity to use short-course records with closer to ten data points.

---

[3] Wikipedia. "List of world records in swimming." https://en.wikipedia.org/wiki/List_ofworldrecordsinswimming (accessed: December 3, 2018).

7.4 Projects

7.4.1 C. Evert

Tennis Network
C. Evert, Commentator
Stamford, CT 06091

Dear Tennis Enthusiasts,

My colleagues and I at the Tennis Network have noticed an interesting trend in the WTA and ATP Grand Slams. We conjecture that players who win the first event of the year, the Australian Open, are more likely to win the last event of the year, the US Open. Of course, both of the events are played on the same surface, which may tend to favor certain players and, further, both events tend to have much larger crowds than the other two Grand Slams. However, this is just speculation and as a reputable journalist I need mathematical support for my conjectured positive trend.

I would like you to determine if there is a correlation between victories in the US Open and victories in the Australian Open for both the men's singles and the women's singles tournaments. Please research the past winners of these tournaments since the Open era began in 1968. Make a scatter plot or scatter plots which compare the number of victories in the US Open with the number of victories in the Australian Open for each player who has won at least one of these tournaments. Write a report describing the findings from your scatter plot and answering the following questions: Is there a correlation between the numbers of victories in these two tournaments? How do you describe the trend? Is it positive or negative, weak or strong, linear or nonlinear? Can you explain the trend or lack of trend? Is there causation from one variable to another?

Sincerely,
C. Evert

7.4.2 C. Maddux

National Baseball Hall of Fame
C. Maddux
25 Main Street
Cooperstown, NY 13326

Dear Introductory-Statistics Student,

I am writing to ask for your help in settling a debate. I have recently taken a job at the National Baseball Hall of Fame in Cooperstown, NY. Some of my, shall we say, more "experienced" colleagues claim that pitching in Major League Baseball is not as good as it was in the golden age 100 years ago. Coming from a family that has supported a more recent player, I claim the complete opposite, and say that in fact pitching is better today than it has been in the past. In order to settle our argument, I am enlisting your help as follows.

Test the hypothesis that today's MLB pitchers are more effective than those of 100 years ago by looking at batting statistics. Compare three statistics: strikeouts per game, walks per game, and hits per game. In 1915, on average, for both the National League and the American League there were 8.2 hits per game, 3.9 strikeouts per game, and 3 walks per game. Research data for MLB teams in the last year. Record these three statistics for both NL and AL teams. Pitchers have different styles, but, in general, pitchers would like strikeouts to go up and walks and hits to go down. Run a hypothesis test to determine if there has been a statistically significant improvement in pitching since 1915. Your final report should include the data collected, the sample statistics, the null and alternative hypotheses for each statistic, the test statistics, and p-values. Conclusions should be drawn and should be supported by the data and analysis.

Thanks again for your help.

Best,

C. Maddux

Implementation Notes This project may seem intimidating for students not familiar with the jargon of baseball. However, baseball statistics are some of the most readily available and clearly displayed statistics of any sport. If you wish to give a little more direction, the ESPN website http://www.espn.com/mlb/statistics gives all the data needed for the current year. Further, if you prefer a narrower focus, we have also utilized this activity by specifying one team and year for comparison.

7.4.3 Chief M. Brody

Amity Island Police Department
Martin Brody, Chief
Amity Island, MA 02535

Dear Shark Expert,

After last night's premiere of Sharknado, my friends and I got into a heated discussion about how to best get away from a hungry shark. Now we know that sharks don't really come flying after you in a tornado, but what if we are off the coast of New England enjoying a swim and a shark fin appears on the horizon? I want to know the fastest way to get away from that shark and back to shore. I say the backstroke is the best; you can glide through the water and watch to see if the shark is catching up. My friend Sam says the butterfly stroke would be far superior, and who wants to be able to watch the shark trying to catch you anyway? A second friend, Matt, claims swimming freestyle is the fastest in the world, while my wife, Ellen, thinks the breaststroke would be best.

Each of these strokes has been a part of the Olympics since 1956. Take samples of Olympic race times to decide which stroke is the fastest. There are many questions to consider when taking a sample. Should you compare the same person racing in all four events? Should you include women and men together or get two separate results? What sampling method would be most practical? Should you compare times for each stroke in the same year or average times across the last 50 years?

Take at least three different samples using three different sampling methods. For comparison, each sample should contain times for all four strokes, and remember, the larger your sample, the more effective your argument CAN be. Then write a report describing how you chose your samples and why each was or was not a representative sample. Utilize a measure of center such as the mean or median to describe a typical time for each stroke within each sample in order to draw conclusions from your data about which stroke would be the best choice for me and each of my three friends.

Thanks,
Martin Brody, Chief

Implementation Notes Some students have difficulty understanding what it means to take three samples. Sometimes they will choose three winning times and consider these samples of size one as effective representatives of the data. To avoid receiving an assignment like this, you may wish to illustrate the point with a specific example when the project is assigned or require rough drafts before the final project is submitted.

8 Probability

This chapter uses professional football, horse racing, and fan bases to investigate topics in probability, including compound and conditional probabilities, expected value, and paradoxes.

8.1 Mathematics

Counting

- Mutually exclusive events
- Independent events

Probabilities

- Probability distribution
- Compound probability
- Conditional probability

Expected Value

- Paradoxes

8.2 Sports

Football Not only is the NFL popular for the football games themselves, but its affiliated activities also garner plenty of attention. Two activities and one project in this chapter will relate to the NFL. The first activity is associated with fantasy football, a game where fans of the sport choose a team of real football players and apply the actual players' statistics in a game against other fans like themselves. Students are expected to use statistics and expected values for Hall of Fame wide receivers to make judgments about who is or was the best fantasy player. The second activity interprets the famous "birthday paradox" in terms of NFL teams. The final project asks students to research outcomes in the NFL draft, that is, the event held annually in late April where college-age students are selected to play on professional teams.

Horse racing While horse racing is not as popular throughout the year as sports such as professional football or soccer, each year the Kentucky Derby receives some of the highest television ratings in the United States. Enthusiasts knowing little about the business of horse racing gather at the track and around their televisions to watch horses and jockeys race out of the gate. The Derby field usually contains over 20 horses racing to win, place, or show, also known as finishing first, second, or third, respectively. With such a large field, the Derby typically runs a wide variety of horses. Probabilities associated with these horses are computed in the activity. With the Kentucky Derby, the Preakness Stakes and the Belmont Stakes form the Triple Crown of horse racing. Horses winning a Triple Crown are unusual, although there have been winners in recent years. The project investigates the historical incidence of Triple Crown winners.

Fans It is the fans of sports that allow sports to exist as they do today. Athletes appreciate the love or disapproval of their fans through applause or boos at the stadiums or arenas and through interactions at events and on social media. The selling of tickets and concessions, television and radio broadcasting rights, and apparel makes professional and college sports alike very big business. In this chapter, fans find their place in an activity and a project as an important part of the sports. Both an activity and a project will investigate probabilities that a person is a fan of a certain sport or team. In the activity the data is given, and students simply calculate basic and compound probabilities. However, in the project, students will gather data based on their own observations before calculating probabilities.

8.3 Activities

8.3.1 Most Popular Sports

Content Probability distribution, independent events, multiplication principle.

Activity Watching sports is a popular pasttime in the United States. A 2015 Harris poll[1] surveyed 2,252 American adults about their favorite sports. Use the table of results below to answer the following questions.

---

[1] The Harris Poll. "Pro Football is Still America's Favorite Sport." (2016) https://theharrispoll.com/\penalty-\@Mnew-york-n-y-this-is-a-conflicting-time-for-football-fans-on-the-one-hand-with-the-big-game-50-no-less-fast-approaching-its-a-time-of-excitement-especial/ (accessed: December 3, 2018).

Sport	Percentage
Pro football	33
Baseball	15
College football	10
Auto racing	6
Men's pro basketball	5
Ice hockey	5
Men's soccer	4
Men's college basketball	4
Men's golf	3
Boxing	3
Other/not sure	12

1. Create an appropriate graphical probability distribution.
2. If you interview five randomly selected Americans, what is the probability that all five say college football is their favorite sport?
3. If you interview six randomly selected Americans, what is the probability that none of the six say baseball is their favorite sport?
4. If you interview four randomly selected Americans, what is the probability that at least one person says men's soccer is his or her favorite sport?
5. Could you answer the previous three questions if your sampling method was not random? Explain.

Implementation Notes Students occasionally get confused by the representation of the probabilities as whole-number percentages and try to calculate a probability by dividing these numbers by the sample size, so this should be made clear if the activity is not being used in a face-to-face setting where the faculty member can correct the mistake.

8.3.2 Stadium Size

Content Independent and mutually exclusive events, conditional probability.

Activity Soccer is easily the most popular sport globally; it is estimated that half the world's population, approximately 4 billion people, consider themselves soccer (association football) fans.[2] Watching sports live in a stadium is an exciting way for

---

[2] Worldatlas. "The Most Popular Sports in the World." https://www.worldatlas.com/articles/what-are-the-most-popular-sports-in-the-world.html (accessed: May 15, 2018).

fans to engage with their favorite teams and spend time with friends and family. The world's eight largest soccer stadiums, measured by capacity, are listed below.[3]

Stadium name	Capacity	Location
Rungrado May Day	150,000	Pyongyang, North Korea
Camp Nou	99,354	Barcelona, Spain
Estadio Azteca	95,500	Mexico City, Mexico
Azadi Stadium	95,225	Tehran, Iran
FNB Stadium	94,736	Johannesburg, South Africa
Rose Bowl	92,542	Pasadena, USA
Wembley Stadium	90,000	London, United Kingdom
Gelora Bung Karno	88,083	Jakarta, Indonesia

1. If you were going to watch an international soccer match in each of these stadiums, in how many different orders is this possible?

2. From each stadium, to how many other stadiums could you drive the entire way to see the next match? From each stadium, what is the probability that you could drive the entire way to see the next match?

3. If you buy your tickets in a random order, what is the probability that you can complete your trip with only one plane ride? What about two, three, or four plane rides?

4. Suppose the first four international matches you attended were at Rungrado May Day, Estadio Azteca, Wembley, and Gelora Bung Karno and, at each match, 25% of the crowd was rooting for the away team. If the home and away teams have an equal chance of winning and a 20% chance of drawing, what is the probability that at least half of the fans at the four games you attended saw their preferred team win?

5. Repeat the last question with the other four stadiums: Camp Nou, Azadi, FNB, and the Rose Bowl. Are your two probabilities similar or different? Considering the relative capacities of the stadiums, is this expected or surprising?

Implementation Notes The initial three questions of this activity are straightforward, although Questions 2 and 3 do require some basic knowledge concerning geography or a map. Questions 4 and 5 are designed to be a bit more challenging; faculty should be prepared to help students take the stadium capacities into account.

Additionally, there are some obvious underpinnings to this activity relating probability to graph theory. Although this activity can be completed independently of that material, if

---

[3] Smith, Matthew Nitch. "The 18 Biggest Soccer Stadiums in the World." Business Insider. May 7, 2016. http://www.businessinsider.com/the-18-biggest-football-stadiums-by-capacity/#10-borg-el-arab-stadium-alexandria-egypt-9 (accessed: May 15, 2018). May not be available outside in US.

graph theory has previously been covered during the semester, creating the representative complete graph and using the language of the traveling salesman problem can be helpful, particularly in the parts concerning modes of transportation. The activity can also easily be lengthened into a project by additionally asking students to create a weighted complete graph and determine efficient routes with varying conditions.

8.3.3 NFL County Paradox

Content Birthday paradox.

Activity This activity is a version of the famous "birthday paradox." There are 53 players on the roster for a team in the National Football League. There are 3,143 counties (or county equivalents) in the United States.

1. Assuming the Denver Broncos have a roster where all 53 players were born in the United States, what is the probability that at least two of the players were born in the same county? (Assume independence among players on the team.)

2. Compare this with the probability that at least one of the players on the Broncos was born in the county where you were born. (If you were not born in the US, choose any single county for comparison.) Explain why these numbers are so different.

3. Now, note that when the Broncos play the Arizona Cardinals, there are 106 people on the two rosters. If all of these players were born in the US, what is the probability that at least two were born in the same county? Is this result surprising? How does this compare with your first calculation?

8.3.4 Kentucky Derby

Content Compound and conditional empirical probabilities, independent events.

Activity The Kentucky Derby is raced every year on the first Saturday in May at Churchill Downs Race Track. The horses compete on a dirt track, with thousands of fans watching from the grandstand and infield as well as millions more watching on television. The mile and a quarter race is sometimes called "the most exciting two minutes in sport." In 2016, the Kentucky Derby had a field of 20 horses. Suppose that you know nothing about horse racing, and so with your limited information each horse is equally likely to win the race. Using the table of 2016 Kentucky Derby horses found in the "Data Sets" appendix (Table A.13), we will explore some unusual ways to pick horses and calculate the empirical probability of winning or losing.

1. "Camptown Races" is a traditional children's song referring to placing a bet based on the color of the horse. What is the probability that a bay horse wins the race?

2. You may decide to pick a horse by a geographic area. What is the probability that the winner is not bred in the Commonwealth of Kentucky?

3. Names are a popular way to choose horses. What is the probability that the winning horse has a two-word name and is chestnut in coloring?

4. What is the probability that the losing horse is a gelding or is ridden by a jockey named Luis?

5. Given that the winning horse is not a colt, what is the probability that the horse is a bay?

In horse racing, a horse may win, place, or show, that is, come in first place, second place, or third place, respectively.

6. In how many ways is it possible for two of the 20 horses to win and place, when we are not interested in the order of the remaining horses? In how many ways is it possible for three of the 20 horses to win, place, and show, again when we are not interested in the order of the remainder?

7. What is the probability that a single trainer has horses that win and place?

8. What is the probability that horses with single-word names win, place, and show?

9. What is the probability that the names of the horses that win, place, and show begin with an "M" if you know that all of the winning horses were bred in Kentucky?

Implementation Notes This activity can help to introduce concepts of probability and counting before providing the students with mathematical formulas.

8.3.5 Best Fantasy Wide Receiver

Content Empirical probabilities, predictions, expected value.

Activity Fantasy football is a popular game where participants earn points based on the actual performance of players in games that week. Wide receivers in particular earn points through touchdowns, receiving yards, and sometimes receptions. In this activity, we will use expected values to determine which player, out of past and present players, we would prefer to have on our fantasy team.

1. Jerry Rice and Terrell Owens are widely regarded as two of the best wide receivers to have played professional football. They lead the league in all-time receiving yards. Rice's career with the 49ers, Raiders, and Seahawks spanned 20 years, while Owens played for 16 years with the Bengals, Bills, Cowboys, Eagles, and 49ers. Rice is a Hall-of-Famer and is called the GOAT, Greatest Of All Time. We will compare his predicted performance with that of Owens using the career statistics given in the

tables below.[4,5] (We note that all statistics are receiving statistics only and do not account for rushing, returning, or passing yards or touchdowns.)

	Years	Games	Receptions	Yards	Touchdowns
Jerry Rice	20	303	1,549	22,895	197

(a) How many receptions would you expect per game for Jerry Rice? How many yards per game? How many touchdowns per game?

(b) For Jerry Rice, over the course of a 16-game season, what is the expected number of receptions? receiving yards? receiving touchdowns?

	Years	Games	Receptions	Yards	Touchdowns
Terrell Owens	16	219	1,078	15,934	154

(c) How many receptions would you expect per game for Terrell Owens? How many yards per game? How many touchdowns per game?

(d) For Terrell Owens, over the course of a 16-game season, what is the expected number of receptions? receiving yards? receiving touchdowns?

(e) Suppose a reception is worth one point, 25 receiving yards are worth one point, and a touchdown is worth six points. Which player would you prefer to have on your team? Support your answer with mathematics.

2. We can argue that career statistics such as total receiving yards are not the best metric for choosing a most valuable player. At the beginning of the 2016 season, some younger wide receivers were putting up impressive records, but have not been and probably will not be able to play for the 20 years that Jerry Rice played. FoxSports.com ranks Antonio Brown and Julio Jones as the best and second best wide receivers, respectively, going in to the 2016–17 football season.[6] Their career statistics are given below.[7,8]

---

[4] NFL. "Jerry Rice." http://www.nfl.com/player/jerryrice/2502642/careerstats (accessed: August 10, 2016).

[5] NFL. "Terrell Owens." http://www.nfl.com/player/terrellowens/2502377/careerstats (accessed: August 10, 2016).

[6] DaSilva, Cameron. "Ranking the Top 10 Wide Receivers in the NFL." Fox Sports. http://www.foxsports.com/nfl/story/nfl-ranking-top-10-wide-receivers-2016-antonio-brown-odell-beckham-julio-jones-062816 (accessed: August 10, 2016).

[7] NFL. "Antonio Brown." http://www.nfl.com/player/antoniobrown/2508061/careerstats (accessed: August 10, 2016).

[8] NFL. "Julio Jones." http://www.nfl.com/player/juliojones/2495454/careerstats (accessed: August 10, 2016).

	Years	Games	Receptions	Yards	Touchdowns
Antonio Brown	6	86	526	7,093	38
Julio Jones	5	65	414	6,201	34

(a) How do these two wide receivers compare in terms of receptions per game, yards per game, and touchdowns per game?

(b) Using the information given here, make an argument for which of the four players you would want on your fantasy team.

(c) What other variables should you consider if you really had to make this decision?

3. Now consider another statistic, called yards per catch.

(a) What is the average yards per catch for Antonio Brown and for Julio Jones over their careers?

(b) Some of the best quarterbacks in NFL history have completion percentages around 65%; that is, 65% of the time they throw a pass, a receiver is able to make the catch. Suppose Antonio Brown and Julio Jones end up on the same team and in one game exactly half of the passes from the quarterback are targeted at each of them. Further, suppose the quarterback typically throws 38 times in the game. Use the yards per catch statistic for Brown and Jones to answer the following.

  i. If the quarterback completes 65% of his passes to each receiver, what is the expected number of receiving yards in this game?

  ii. If the quarterback completes 80% of his passes to Jones but only 50% to Brown, what is the expected number of receiving yards in this game?

Implementation Notes Question 3 is easily omitted if you are constrained by time.

8.4 Projects

8.4.1 B. Oaks

At the Post Stables
Brittany Oaks, Owner
Paris, KY 40361

Dear Student Researcher,

I am the new owner of At the Post Stables and am looking for your help. After I won the lottery last year, I thought owning a thoroughbred stable would be such a kick, but to be perfectly honest, I am in over my head in the horse world. There are many things that I don't understand, but, in particular, I would like you to look into the Triple Crown for me. I know the Triple Crown consists of three stakes races, the Kentucky Derby, the Preakness Stakes, and the Belmont Stakes, which have collectively been raced since 1875. I'd like to be more informed on the probabilities of a horse successfully winning these three races, so first explain the difference between empirical and theoretical probabilities. Then do some research to find out which horses have won the Triple Crown since 1875 and use your data to answer these questions.

- What is the empirical probability of having a Triple Crown winner this year?
- What is the empirical probability of at least one Triple Crown in the next 20 years?
- How many years will it be until having at least one Triple Crown winner is more likely than not?
- Hypothetically, if each of the three stakes race the same 20 horses, what is the theoretical probability of a Triple Crown winner in a given year?

To add to my confusion, I read an article written before the 2015 race which speculated that the Triple Crown would never be completed again.[9] However, since that time two horses, American Pharaoh in 2015 and Justify in 2018, have won the Triple Crown. Create a graph illustrating years between Triple Crown winners.

- What can you say about the trend of the graph?
- Find the mean and standard deviation of the number of years between Triple Crown winners.
- Interpret the mean in terms of a probability. How many years should pass before it's more likely than not to have another Triple Crown winner? Please explain.

Good luck with your research!

Sincerely,
Brittany Oaks
At the Post Stables

8.4.2 Jock Jones

National Sports Apparel Company
Jock Jones, Design Manager
Beaverton, OR 97005

[9] Morris, Benjamin. "Don't Expect to See a Triple Crown Anytime Soon." FiveThirtyEight. June 9, 2014. http://fivethirtyeight.com/datalab/dont-expect-to-see-a-triple-crown-anytime-soon/ (accessed: August 10, 2016).

Dear Observer,

People love to support their favorite sports teams through their clothing. We at the National Sports Apparel Company need to know about the trending teams in your area. Our job for you is this: Choose a public place where a large number of people will probably go past you, like a park, grocery store, movie theater, or mall, and spend some time people-watching. For each person walking past who is wearing an item of sports apparel, record the team supported and whether the person is male or female. Collect 100 data points, and then create a two-way table summarizing the data by team and gender.

Use your table to produce an empirical probability distribution and to report the interesting characteristics of your data, answering questions such as: What team is the most popular? If there are 500 fans, how many would you expect to belong to each fanbase? What proportion of the sample was men? Is that amount what you expected? What are some of the differences between the distributions of sports teams for men and women? What is the probability that a person is a fan of the most popular team given that she is a woman, or given that he is a man? Conclude your report with your recommendations on which teams we should focus on in the apparel design for your area and whether the marketing plan should be different for women and men.

Thank you for your help in this matter.

Sincerely,
Jock Jones
National Sports Apparel Company

Implementation Notes If you expect the data to be particularly biased toward a team in your area, you may want to specify a particular professional sport or college sport to avoid that team and create more diversity in the results. However, this does make gathering the sample more time-consuming.

8.4.3 A Football Mom

A Football Mom
Tuscaloosa, AL 35487

Dear Draft Analyst,

Every year in April, the National Football League Player Selection Meeting, also known as the NFL draft, takes place. Over seven rounds, team owners select eligible college players, hoping to improve their team for the coming years. As it turns out, my son is demonstrating a talent for football that I would like to encourage. It is well known that higher draft picks will receive higher salaries, with more guaranteed money than lower picks. My son and I are aiming for him to be picked in the first round. With that in mind, we need to plan now to choose his position and school in order to maximize his chances of being taken in the first round.

Research the last five years of NFL drafts to determine how many of each year's 160 players played each position. Which position is most likely to be drafted? Is it the most likely by a substantial margin? Are offensive or defensive players more likely to be chosen? What position would you advise my son to focus on and why?

Over this same time period, determine how many athletes chose to play college football and, for those that did, record either the conference where each player chose to play or the community college if they did not play in Division I. Based on this empirical data, which conference is most likely to provide a first-round draft pick? If my son chooses to go to school in this conference, does that change your recommendation for which position will most likely get him drafted? Further, given the position initially recommended above, which conference would be the best choice?

Please provide both your data and your conclusions. We look forward to your response.

Sincerely,
A Football Mom

Data Sets

Table A.1 NBA playoff bonus pools($)–2010–2011 season. Data for activity in Section 5.3.4

Team	Top record	Conference record	Round 1	Conference semifinals	Conference finals	Finals	Total
Mavs		212,559	179,092	213,095	352,137	2,125,137	3,082,020
Heat		243,411	179,092	213,095	352,137	1,408,168	2,395,903
Bulls	346,105	302,841	179,092	213,095	352,137		1,393,270
Thunder		142,800	179,092	213,095	352,137		887,124
Lakers		212,559	179,092	213,095			604,746
Celtics		181,706	179,092	213,095			573,893
Hawks		118,990	179,092	213,095			511,177
Spurs		302,841	179,092				481,933
Grizzlies			179,092	213,095			392,187
Magic		142,800	179,092				321,892
Nuggets		118,990	179,092				298,082
Knicks		81,157	179,092				260,249
Blazers		81,157	179,092				260,249
76ers			179,092				179,092
Pacers			179,092				179,092
Hornets			179,092				179,092
Total							12,000,000

Albert. "NBA Championship Prize Money." Heat Hoops.com. http://heathoops.com/2013/06/2013-nba-championship-prize-money/ (accessed: July 22, 2016).

Table A.2 NBA playoff bonus pools ($)—2015–2016 seasons. Data for activity in Section 5.3.4

Team	Top record	Conference record	Round 1	Conference semifinals	Conference finals	Finals	Total
Cavaliers	378,553		223,864	266,369	440,173	2,656,422	3,965,381
Warriors	432,632	378,553	223,864	266,369	440,173	1,760,210	3,501,801
Thunder	227,132		223,864	266,369	440,173		1,157,538
Raptors	304,263		223,864	266,369	440,173		1,234,669
Spurs	304,263		223,864	266,369			794,496
Heat	163,955		223,864	266,369			654,188
Hawks	163,955		223,864	266,369			654,188
Blazers	148,738		223,864	266,369			638,971
Clippers	178,501		223,864				402,365
Celtics	163,955		223,864				387,819
Hornets	163,955		223,864				387,819
Mavericks	50,724		223,864				274,588
Grizzlies	50,724		223,864				274,588
Pacers			223,864				223,864
Pistons			223,864				223,864
Rockets			223,864				223,864
Total							15,000,000

Table A.3 Annual returns—Dow Jones. Data for activity in Section 5.3.4

Year	Gain or loss (%)
2010	11.02
2011	5.53
2012	7.26
2013	26.50
2014	7.52
2015	−2.23
2016	13.42
2017	25.08

Macrotrends. "Dow Jones - 100 Year Historical Chart." https://www.macrotrends.net/1319/dow-jones-100-year-historical-chart (accessed: December 3, 2018).

Table A.4 Annual returns—S&P 500. Data for activity in Section 5.3.4

Year	Gain or loss (%)
2010	12.78
2011	0.00
2012	13.41
2013	29.60
2014	11.39
2015	−0.73
2016	9.54
2017	19.42

Macrotrends. "S&P 500 Historical Annual Returns." https://www.macrotrends.net/2526/sp-500-historical-annual-returns (accessed: December 3, 2018).

Table A.5 Boston Marathon—men's championship times. Data for activity in Section 6.3.1

Year	Name	Country	Time
1980	Bill Rodgers	USA	2:12:11
1981	Toshihiko Seko	JPN	2:09:26
1982	Alberto Salazar	USA	2:08:52
1983	Greg Meyer	USA	2:09:00
1984	Geoff Smith	GBR	2:10:34
1985	Geoff Smith	GBR	2:14:05
1986	Robert de Castella	AUS	2:07:51
1987	Toshihiko Seko	JPN	2:11:50
1988	Ibrahim Hussein	KEN	2:08:43
1989	Abebe Mekonnen	ETH	2:09:06
1990	Gelindo Bordin	ITA	2:08:19
1991	Ibrahim Hussein	KEN	2:11:06
1992	Ibrahim Hussein	KEN	2:08:14
1993	Cosmas Ndeti	KEN	2:09:33
1994	Cosmas Ndeti	KEN	2:07:15
1995	Cosmas Ndeti	KEN	2:09:22
1996	Moses Tanui	KEN	2:09:15
1997	Lameck Aguta	KEN	2:10:34
1998	Moses Tanui	KEN	2:07:34
1999	Joseph Chebet	KEN	2:09:52
2000	Elijah Lagat	KEN	2:09:47
2001	Lee Bong-Ju	KOR	2:09:43
2002	Rodgers Rop	KEN	2:09:02
2003	Robert Kipkoech Cheruiyot	KEN	2:10:11
2004	Timothy Cherigat	KEN	2:10:37
2005	Hailu Negussie	ETH	2:11:45
2006	Robert Kipkoech Cheruiyot	KEN	2:07:14
2007	Robert Kipkoech Cheruiyot	KEN	2:14:13
2008	Robert Kipkoech Cheruiyot	KEN	2:07:46

Table A.5 Continued

Year	Name	Country	Time
2009	Deriba Merga	ETH	2:08:42
2010	Robert Kiprono Cheruiyot	KEN	2:05:52
2011	Geoffrey Mutai	KEN	2:03:02
2012	Wesley Korir	KEN	2:12:40
2013	Lelisa Desisa	ETH	2:10:22
2014	Meb Keflezighi	USA	2:08:37

Boston Athletic Association. "Champions of the Boston Marathon: Men's Open Division." https://www.baa.org/races/boston-marathon/results/champions (accessed: December 3, 2018).

Table A.6 2013 Red Sox hitting statistics. Data for activity in Section 6.3.3

Name	AB	BB	H	S	2B	3B	HR	TB
Stephen Drew	442	54	112	62	29	8	13	196
Jacoby Ellsbury	577	47	172	124	31	8	9	246
Jonny Gomes	312	43	77	47	17	0	13	133
Mike Napoli	498	73	129	66	38	2	23	240
Daniel Nava	458	51	139	98	29	0	12	204
David Ortiz	518	76	160	90	38	2	30	292
Dustin Pedroia	641	73	193	140	42	2	9	266
Jarrod Saltalamacchia	425	43	116	62	40	0	14	198
Shane Victorino	477	25	140	97	26	2	15	215

ESPN. "Boston Red Sox Batting Stats—2013." http://www.espn.com/mlb/team/stats/batting/_/name/bos/year/2013 (accessed: December 3, 2018).

The following baseball acronyms and terminology are used in the activity in Section 6.3.3.

AB *At bats* is the total number of times that a player either got a hit or made an out.

H *Hits* is the total number of times a player reached base by getting a single, double, triple, or home run.

S *Singles* is the total number of times a player reached base by getting to first base.

2B *Doubles* is the total number of time a player reached base by getting to second base.

3B *Triples* is the total number of time a player reached base by getting to third base.

HR *Home runs* is the total number of time a player reached base by getting to home plate.

TB   *Total bases* is the sum of the numbers, of bases that a player reached. In this calculation, first base counts as one base, second base counts as two bases, third base counts as three, and home plate as four.

BB   *Base-on-balls*, or *walks*, is the number of times that a player reached first base without getting a hit due to the pitcher missing the strike zone four times in a plate appearance.

PA   *Plate appearances* is the sum of AB and BB. Since the event of a walk does not increase a players, AB, plate appearances count the total number of times a batter had the opportunity to get a hit or make an out.

OBP   *On-base percentage* is the ratio of the number of times a player reached base to his plate appearances.

SLG   *Slugging percentage* is a weighted average of a batter's hits to his at bats. The weighting counts singles as 1, doubles as 2, triples as 3, and home runs as 4; thus it essentially is a ratio of total bases to at bats.

Table A.7 2014 Masters—putts per hole. Data for activity in Section 6.3.5

Player	Round 1	Round 2	Round 3	Round 4	Total
L. Donald	1.67	1.22			1.44
R. Fowler	1.44	1.56	1.50	1.50	1.50
C. Lee	1.61	1.39			1.50
C. Schwartzel	1.33	1.67			1.50
R. Henley	1.56	1.33	1.61	1.56	1.51
J. Blixt	1.50	1.61	1.50	1.50	1.53
P. Hanson	1.56	1.50			1.53
I. Poulter	1.67	1.44	1.50	1.56	1.54
S. Bowditch	1.61	1.61	1.61	1.39	1.56
S. Stricker	1.44	1.56	1.50	1.72	1.56
L. Mize	1.56	1.28	1.67	1.72	1.56
E. Els	1.44	1.67			1.56
M. Jimenez	1.56	1.67	1.39	1.61	1.56
M. Weir	1.61	1.50	1.78	1.39	1.57
V. Singh	1.50	1.67	1.61	1.56	1.58
W. Simpson	1.67	1.50			1.58

https://www.masters.com/en_US/scores/stats/putts.html (accessed: 2014). Note: The Masters website is updated yearly and only shows data for the most recent Masters tournament.

Player	Round 1	Round 2	Round 3	Round 4	Total
D. Clarke	1.72	1.50	1.56	1.56	1.58
T. Bjorn	1.78	1.50	1.56	1.50	1.58
C. Kirk	1.67	1.39	1.72	1.61	1.60
G. Fernández-Castaño	1.61	1.39	1.67	1.72	1.60
G. Woodland	1.50	1.83	1.44	1.61	1.60
M. Leishman	1.50	1.72			1.61
K. Stadler	1.72	1.67	1.67	1.39	1.61
B. Langer	1.61	1.72	1.78	1.33	1.61
J. Olazabal	1.50	1.56	1.72	1.67	1.61
B. de Jonge	1.61	1.50	1.78	1.56	1.61
B. Watson	1.78	1.44	1.83	1.39	1.61
M. Kuchar	1.72	1.67	1.39	1.67	1.61
M. Fitzpatrick	1.61	1.61			1.61
D. Ernst	1.61	1.61			1.61
B. Grace	1.67	1.56			1.61
Z. Johnson	1.72	1.50			1.61
B. Snedeker	1.44	1.50	1.94	1.61	1.62
J. Senden	1.72	1.50	1.61	1.67	1.62
J. Donaldson	1.56	1.50	1.78	1.67	1.62
S. Gallacher	1.72	1.56	1.89	1.33	1.62
Y. Yang	1.78	1.50			1.64
D. Lynn	1.67	1.61			1.64
K. Bradley	1.72	1.56			1.64
S. Lyle	1.61	1.56	1.83	1.61	1.65
K. Streelman	1.50	1.61	1.67	1.83	1.65
F. Couples	1.72	1.61	1.72	1.56	1.65
L. Westwood	1.78	1.61	1.50	1.72	1.65
L. Oosthuizen	1.56	1.78	1.78	1.56	1.67
H. Mahan	1.67	1.67	1.61	1.72	1.67
S. Bae	1.56	1.78			1.67
B. Crenshaw	1.72	1.61			1.67
M. Kaymer	1.67	1.83	1.50	1.67	1.67

(*Continued*)

Table A.7 Continued

Player	Round 1	Round 2	Round 3	Round 4	Total
R. Moore	1.72	1.61			1.67
H. English	1.67	1.67			1.67
J. Spieth	1.78	1.67	1.61	1.61	1.67
J. Rose	1.61	1.56	1.78	1.72	1.67
J. Furyk	1.61	1.56	1.67	1.83	1.67
B. Haas	1.56	1.67	1.89	1.61	1.68
L. Glover	1.67	1.50	1.78	1.78	1.68
N. Watney	1.50	1.67	1.94	1.61	1.68
T. Jaidee	1.78	1.78	1.39	1.78	1.68
J. Day	1.78	1.61	1.56	1.78	1.68
K. Choi	1.50	1.83	1.83	1.56	1.68
A. Scott	1.67	1.56	1.94	1.61	1.69
T. Olesen	1.83	1.61	1.83	1.50	1.69
O. Goss	1.78	1.56	1.61	1.83	1.69
F. Molinari	1.67	1.72	1.67	1.72	1.69
K. Duke	1.67	1.72			1.69
J. Huh	1.67	1.72			1.69
C. Stadler	1.94	1.44			1.69
J. Luiten	1.78	1.67	1.83	1.56	1.71
T. Watson	1.83	1.61			1.72
S. Cink	1.56	1.72	1.83	1.78	1.72
H. Stenson	1.78	1.78	1.83	1.50	1.72
T. Clark	1.67	1.78			1.72
D. Johnson	1.72	1.72			1.72
M. Jones	1.61	1.83			1.72
P. Mickelson	1.78	1.67			1.72
R. Mcilroy	1.89	1.72	1.67	1.67	1.74
V. Dubuisson	1.72	1.78			1.75
A. Cabrera	1.83	1.67			1.75
M. Every	1.83	1.67			1.75
S. Stallings	1.89	1.61			1.75
J. Walker	1.72	1.83	1.89	1.56	1.75
S. Garcia	1.72	1.83			1.78

Player	Round 1	Round 2	Round 3	Round 4	Total
J. Dufner	1.89	1.67			1.78
D. Points	1.94	1.61			1.78
B. Weekley	1.72	1.83			1.78
M. O'Meara	1.72	1.89			1.81
G. McDowell	1.78	1.83			1.81
I. Woosnam	1.89	1.72			1.81
B. Horschel	1.83	1.94	1.89	1.61	1.82
P. Reed	1.72	1.94			1.83
M. McCcoy	1.78	1.89			1.83
T. Immelman	1.89	1.78			1.83
M. Manassero	1.78	1.94			1.86
G. Porteous	1.72	2.00			1.86
J. Niebrugge	2.00	1.72			1.86
R. Castro	1.89	1.83			1.86
H. Matsuyama	2.17	1.67			1.92
G. DeLaet	2.17	1.72			1.94

Table A.8 2014 Masters—birdies. Data for activity in Section 6.3.5

Player	Round 1	Round 2	Round 3	Round 4	Total
J. Senden	3	6	4	5	18
J. Walker	6	3	2	6	17
M. Jimenez	4	1	7	5	17
K. Stadler	4	4	3	5	16
S. Gallacher	4	4	2	6	16
R. McIlroy	4	2	4	6	16
J. Blixt	6	4	4	2	16
C. Kirk	3	5	4	4	16
A. Scott	5	3	2	5	15
J. Olazabal	4	2	3	6	15
B. Haas	6	3	2	4	15

https://www.masters.com/en_US/scores/stats/birdies.html (accessed: 2014).
Note: The Master website is updated yearly and only shows data for the most recent Masters tournament.

(*Continued*)

Table A.8 Continued

Player	Round 1	Round 2	Round 3	Round 4	Total
J. Donaldson	4	4	2	5	15
B. Watson	3	6	1	5	15
M. Kuchar	2	4	6	3	15
J. Day	2	4	4	5	15
B. Langer	4	2	2	6	14
L. Glover	4	6	1	3	14
J. Furyk	4	4	3	3	14
J. Spieth	3	3	4	4	14
F. Couples	3	4	4	3	14
T. Bjorn	2	8	2	2	14
J. Rose	3	4	2	5	14
M. Weir	3	4	2	4	13
B. Snedeker	5	2	3	3	13
R. Fowler	3	2	6	2	13
I. Poulter	1	5	5	2	13
D. Clarke	3	3	3	4	13
N. Watney	4	2	3	4	13
T. Olesen	2	4	3	4	13
T. Jaidee	4	2	2	5	13
B. Horschel	2	4	2	4	12
K. Choi	3	3	1	5	12
L. Oosthuizen	6	1	2	3	12
B. de Jonge	3	5	1	3	12
G. Fernández-Castaño	1	6	1	4	12
S. Stricker	4	2	3	3	12
R. Henley	1	5	4	2	12
J. Luiten	1	3	2	6	12
G. Woodland	3	2	6	1	12
L. Westwood	2	2	4	4	12

Player	Round 1	Round 2	Round 3	Round 4	Total
H. Stenson	2	3	1	5	11
S. Bowditch	1	3	3	4	11
S. Cink	2	3	2	4	11
S. Lyle	3	5	1	2	11
V. Singh	2	3	3	2	10
M. Kaymer	3	3	3	1	10
K. Streelman	3	3	3	1	10
H. Mahan	1	2	3	3	9
F. Molinari	4		2	3	9
L. Donald	3	5			8
O. Goss	2	3	2	1	8
R. Moore	3	5			8
W. Simpson	3	5			8
L. Mize	2	3	2	1	8
H. Matsuyama	2	5			7
M. Leishman	4	3			7
P. Reed	3	3			6
P. Mickelson	2	4			6
S. Garcia	4	2			6
C. Lee	2	4			6
D. Points	2	4			6
K. Bradley	3	3			6
V. Dubuisson	4	1			5
D. Lynn	2	3			5
J. Huh	2	3			5
G. DeLaet		5			5
B. Grace	1	4			5
E. Els	3	2			5

(*Continued*)

Table A.8 Continued

Player	Round 1	Round 2	Round 3	Round 4	Total
T. Immelman	2	3			5
T. Watson	3	2			5
A. Cabrera	2	3			5
D. Ernst	3	2			5
M. Every	2	3			5
I. Woosnam	2	3			5
S. Stallings	1	3			4
G. McDowell	2	2			4
J. Dufner	1	3			4
M. Manassero	4				4
C. Stadler	1	3			4
G. Porteous	2	2			4
S. Bae	1	3			4
H. English	3	1			4
B. Weekley	3	1			4
C. Schwartzel	3	1			4
P. Hanson		3			3
D. Johnson	2	1			3
Y. Yang	1	2			3
Z. Johnson		3			3
M. McCoy	2	1			3
M. Fitzpatrick	1	2			3
T. Clark	1	2			3
R. Castro	2	1			3
K. Duke	1	1			2
B. Crenshaw	1	1			2
M. Jones	2				2
J. Niebrugge		2			2
M. O'Meara		1			1

Table A.9 Complete games by day in the 2013 MLB regular season. Data for activity in Section 7.3.1

Week	Sunday	Monday	Tuesday	Wednesday	Thursday	Friday	Saturday
1	Mar 31	Apr 1	Apr 2	Apr 3	Apr 4	Apr 5	Apr 6
(Days 1–7)	0	1	0	0	0	0	0
2	Apr 7	Apr 8	Apr 9	Apr 10	Apr 11	Apr 12	Apr 13
(Days 8–14)	0	0	0	1	0	1	1
3	Apr 14	Apr 15	Apr 16	Apr 17	Apr 18	Apr 19	Apr 20
(Days 9–21)	2	1	1	0	0	0	1
4	Apr 21	Apr 22	Apr 23	Apr 24	Apr 25	Apr 26	Apr 27
(Days 22–28)	0	0	1	0	0	2	0
5	Apr 28	Apr 29	Apr 30	May 1	May 2	May 3	May 4
(Days 29–35)	1	1	0	1	0	1	1
6	May 5	May 6	May 7	May 8	May 9	May 10	May 11
(Days 36–42)	0	0	0	0	0	2	1
7	May 12	May 13	May 14	May 15	May 16	May 17	May 18
(Days 43–49)	1	1	0	0	0	0	2
8	May 19	May 20	May 21	May 22	May 23	May 24	May 25
(Days 50–56)	0	2	0	1	0	1	1
9	May 26	May 27	May 28	May 29	May 30	May 31	Jun 1
(Days 57–63)	0	2	1	0	0	1	1
10	Jun 2	Jun 3	Jun 4	Jun 5	Jun 6	Jun 7	Jun 8
(Days 64–70)	1	1	0	1	0	0	0
11	Jun 9	Jun 10	Jun 11	Jun 12	Jun 13	Jun 14	Jun 15
(Days 71–77)	0	1	1	0	0	2	0
12	Jun 16	Jun 17	Jun 18	Jun 19	Jun 20	Jun 21	Jun 22
(Days 78–84)	0	1	0	1	0	0	0
13	Jun 23	Jun 24	Jun 25	Jun 26	Jun 27	Jun 28	Jul 29
(Days 85–91)	0	0	0	2	1	0	2

Retrosheet. "Game Logs for Individual Seasons." https://www.retrosheet.org/gamelogs/ (accessed: December 3, 2018).

(Continued)

Table A.9 Continued

Week	Sunday	Monday	Tuesday	Wednesday	Thursday	Friday	Saturday
14	Jun 30	Jul 1	Jul 2	Jul 3	Jul 4	Jul 5	Jul 6
(Days 92–98)	1	1	2	0	0	2	0
15	Jul 7	Jul 8	Jul 9	Jul 10	Jul 11	Jul 12	Jul 13
(Days 99–105)	1	0	3	0	0	1	3
16	Jul 14	Jul 15	Jul 16	Jul 17	Jul 18	Jul 19	Jul 20
(Days 106–112)	1	0	0	0	0	0	0
17	Jul 21	Jul 22	Jul 23	Jul 24	Jul 25	Jul 26	Jul 27
(Days 113–119)	2	2	0	1	1	1	2
18	Jul 28	Jul 29	Jul 30	Jul 31	Aug 1	Aug 2	Aug 3
(Days 120–126)	0	0	1	0	0	0	1
19	Aug 4	Aug 5	Aug 6	Aug 7	Aug 8	Aug 9	Aug 10
(Days 127–133)	1	1	0	1	0	0	0
20	Aug 11	Aug 12	Aug 13	Aug 14	Aug 15	Aug 16	Aug 17
(Days 134–140)	3	4	0	1	0	0	0
21	Aug 18	Aug 19	Aug 20	Aug 21	Aug 22	Aug 23	Aug 24
(Days 141–147)	0	2	1	1	0	2	0
22	Aug 25	Aug 26	Aug 27	Aug 28	Aug 29	Aug 30	Aug 31
(Days 148–154)	1	0	2	0	0	0	2
23	Sep 1	Sep 2	Sep 3	Sep 4	Sep 5	Sep 6	Sep 7
(Days 155–161)	1	2	0	0	0	3	0
24	Sep 8	Sep 9	Sep 10	Sep 11	Sep 12	Sep 13	Sep 14
(Days 162–168)	0	1	1	0	0	1	1
25	Sep 15	Sep 16	Sep 17	Sep 18	Sep 19	Sep 20	Sep 21
(Days 169–175)	1	2	0	0	1	2	2
26	Sep 22	Sep 23	Sep 24	Sep 25	Sep 26	Sep 27	Sep 28
(Days 176–182)	1	1	1	1	0	1	1
27	Sep 29	Sep 30					
(Days 183–184)	1	1					

Table A.10 Games per match (men's singles) at Wimbledon in 2014. Data for activity in Section 7.3.2

Row	Round					
1	1	Djokovic v. Golubev	23	Kamke v. Hernych	59	
2	1	Brown v. Baghdatis	49	Anderson v. Bedene	35	
3	1	Chardy v. Cox	50	Haase v. Pospisil	50	
4	1	Youzhny v. Ward	27	Seppi v. Mayer	54	
5	1	Johnson v. Bautista Agut	46	Gabashvili v. Puetz	40	
6	1	Groth v. Dologopolov	44	Stepanek v. Cuevas	33	
7	1	Kravchuk v. Simon	37	Hanescu v. Berdych	45	
8	1	Matosevic v. Verdasco	44	Dimitrov v. Harrison	33	
9	1	Saville v. Thiem	47	Haider-Maurer v. Edmund	36	
10	1	Cilic v. Mathieu	42	Kuznetsov v. Fognini	51	
11	1	Kuznetsov (Andrey) v. Evans	42	Carreno Busta v. Ferrer	37	
12	1	Volandri v. Roger-Vasselin	35	Young v. Becker	35	
13	1	Tomic v. Donskoy	33	Wang v. Gonzalez	32	
14	1	Stakhovsky v. Berlocq	34	Gulbis v. Zopp	42	
15	1	Andujar v. Rola	33	Murray v. Goffin	32	
16	1	Klizan v. Nadal	43	Herbert v. Sock	51	
17	1	Lopez v. Sugita	41	Lacko v. Robredo	42	
18	1	Jaziri v. Monfils	38	Istomin v. Tursunov	43	
19	1	Smethurst v. Isner	35	Riba v. Mannarino	34	
20	1	Nieminen v. Delbonis	39	Falla v. Pavic	35	
21	1	Sela v. Kukushkin	36	Granollers v. Mahut	51	
22	1	De Schepper v. Nishikori	41	Janowicz v. Devvarman	52	
23	1	Hewitt v. Przysiezny	41	Lajovic v. Garcia-Lopez	54	
24	1	Ilhan v. Kudla	48	Lorenzi v. Federer	30	
25	1	Muller v. Benneteau	42	Kohlschrieber v. Sijsling	31	
26	1	Gasquet v. Duckworth	50	Karlovic v. Dancevic	42	
27	1	Raonic v. Ebden	29	Klahn v. Querrey	48	
28	1	Paire v. Rosol	48	Russell v. Reister	61	
29	1	Ito v. Bolelli	47	Vesely v. Burgos	12	
30	1	Lu v. Nedovyesov	50	Kyrgios v. Robert	53	
31	1	Wawrinka v. Sousa	34	Kubot v. Struff	33	

From data collected by IBM and Wimbledon during the 2014 championship (accessed: June 23–July 6, 2014). Results may also be found on ESPN.com at http://www.espn.com/tennis/.

(*Continued*)

Table A.10 Continued

Row	Round					
32	1	Gimeno-Traver v. Giraldo	31	Melzer v. Tsonga		43
33	2	Dimitrov v. Saville	27	Tomic v. Berdych		43
34	2	Puetz v. Fognini	40	Youzhny v. Wang		43
35	2	Kuznetsov v. Ferrer	45	Mayer v. Bahdatis		40
36	2	Gulbis v. Stakhovsky	32	Anderson v. Roger-Vasselin		39
37	2	Djokovic v. Stepanek	45	Cilic v. Haider-Maurer		36
38	2	Simon v. Haase	33	Hernych v. Bautista Agut		38
39	2	Querrey v. Tsonga	71	Becker v. Dolgopolov		49
40	2	Chardy v. Matosevic	61	Murray v. Rola		20
41	2	Wawrinka v. Lu	43	Mannarino v.Robredo		30
42	2	Nieminen v. Isner	38	Dancevic v. Kukushkin		26
43	2	Kohlschrieber v. Bolelli	49	Gasquet v. Kyrgios		62
44	2	Kubot v. Lajovic	48	Muller v. Federer		30
45	2	Lopez v. Pavic	35	Granollers v. Giraldo		49
46	2	Janowicz v. Hewitt	22	Reister v. Istomin		33
47	2	Vesely v. Monfils	58	Kudla v. Nishikori		24
48	2	Raonic v. Sock	29	Rosol v. Nadal		43
49	3	Murray v. Bautistia Agut	25	Stakhovsky v. Chardy		37
50	3	Mayer v. Kuznetsov	32	Wang v. Tsonga		28
51	3	Janowicz v. Robredo	50	Djokovic v. Simon		28
52	3	Dimitrov v. Dogopolov	48	Wawrinka v. Istomin		43
53	3	Lopez v. Isner	35	Cilic v. Berdych		36
54	3	Anderson v. Fognini	43	Giraldo v. Federer		25
55	3	Kyrgios v. Vesely	38	Kukushkin v. Nadal		34
56	3	Raonic v. Kubot	34	Bolelli v. Nishikori		24
57	4	Murray v. Anderson	32	Dimitrov v. Mayer		31
58	4	Wawrinka v. Lopez	35	Raonic v. Nishikori		39
59	4	Djokovic v. Tsonga	31	Chardy v. Cilic		33
60	4	Robredo v. Federer	27	Kyrgios v. Nadal		49
61	Q	Murray v. Dimitrov	28	Djokovic v. Cilic		45
62	Q	Warinka v. Federer	42	Raonic v. Kyrgios		44
63	S	Federer v. Raonic	30	Djokovic v. Dimitrov		36
64	F	Djokovic v. Federer	58			

Table A.11 Unforced errors per hour in the 2014 French Open. Data for activity in Section 7.3.3

Data	Player	UEpH	Player	UEpH	Player	UEpH
1–3	Sousa	12.55	Wawrinka	26.01	Pavic	23.62
4–6	Djokovic	18.00	Garcia-Lopez	11.75	Simon	14.04
7–9	Nadal	8.82	Klizan	13.85	Bagnis	14.83
10–12	Ginepri	24.12	Nishikori	20.51	Benneteau	11.01
13–15	Mahut	17.14	Mathieu	12.66	Montanes	6.00
16–18	Kukushkin	14.08	Thiem	12.66	De Schepper	9.00
19–21	Goffin	14.78	Kubot	13.33	Mecir	18.68
22–24	Melzer	10.72	Gulbis	11.11	Kamke	11.13
25–27	Duckworth	11.77	Becker	15.09	Robredo	11.79
28–30	Mayer	9.87	Bellucci	13.71	Ward	15.95
31–33	Falla	14.69	Beck	21.39	Dodig	21.82
34–36	Paire	26.33	Fognini	16.63	Granollers	38.18
37–39	Gabashvili	18.60	Russell	23.69	Young	19.70
40–42	Pospisil	33.00	Gonzalez	13.38	Sela	26.42
43–45	Rosol	21.43	Lorenzi	10.79	Hasse	8.00
46–48	Vesely	12.38	Bautista Agut	12.09	Dvydenko	22.50
49–51	Dzumhur	17.65	Cilic	16.95	Mannarino	7.30
52–54	Lopez	17.65	Andujar	10.53	Lu	17.84
55–57	Lacko	15.00	Tsonga	12.77	Raonic	15.00
58–60	Federer	13.57	Roger–Vasselin	11.06	Kyrgios	14.50
61–63	Herbert	6.27	Berdych	18.31	Chardy	19.79
64–66	Isner	14.12	Polansky	10.68	Gimeno–Traver	23.59
67–69	Youzhny	14.75	Volandri	13.45	Nieminen	13.95
70–72	Carreno Busta	9.39	Querrey	22.76	Przysiezny	21.05
73–75	Tursunov	8.57	Dolgopolov	11.54	Stepanek	19.62
76–78	Starace	9.71	Ramos	8.31	Arguello	20.48
79–81	Estrella Burgos	18.46	Devvarman	9.64	Elias	19.58
82–84	Janowicz	28.11	Nedovyesov	18.96	Schwartzman	19.17
85–87	Sijsling	31.80	Monfils	18.68	Gasquet	7.42
88–90	Ferrer	12.60	Hanescu	11.13	Tomic	22.27

From data collected by IBM and Roland Garros during the 2014 championship (accessed: May 5–June 8, 2014). Results may be also found on ESPN.com http://www.espn.com/tennis/.

(Continued)

Table A.11 Continued

Data	Player	UEpH	Player	UEpH	Player	UEpH
91–93	Golubev	22.84	Dimitrov	11.19	Llodra	14.21
94–96	Murray	13.16	Karlovic	8.14	Verdasco	7.11
97–99	Monaco	10.80	Olivetti	21.98	Kohlschrieber	14.58
100–102	Pouille	25.40	Struff	5.35	Riba	19.07
103–105	Zopp	8.89	Starkovsky	9.93	Haider–Maurer	12.50
106–108	Haas	11.11	Istomin	11.22	Brands	13.21
109–111	Matosevic	8.62	Robert	16.36	Almagro	28.57
112–114	Brown	16.24	Anderson	14.73	Sock	11.43
115–117	Hewitt	16.91	Bolelli	20.32	Ebden	25.50
118–120	Berlocq	8.30	Arnaboldi	13.55	Cuevas	11.25
121–123	Lajovic	14.05	Seppi	18.92	Michon	18.05
124–126	Delbonis	24.32	Giraldo	30.81	Klahn	24.69
127–129	Johnson	11.39	Chardy	22.83	Tsonga	11.21
130–132	Lokoli	15.44	Djokovic	11.74	Melzer	16.82
133–135	Gonzalez	12.12	Schwartzman	20.97	Paire	30.92
136–138	Simon	19.04	Federer	19.81	Baustista Agut	7.38
139–141	Berdych	11.38	Bagnis	22.05	Robredo	3.56
142–144	Nedovyesov	14.83	Gulbis	11.79	De Schepper	22.57
145–147	Youzhny	14.47	Raonic	11.72	Tursunov	13.07
148–150	Stepanek	7.66	Vesely	7.97	Querrey	19.60
151–153	Haase	11.35	Kukushkin	8.79	Cilic	15.74
154–156	Klizan	21.08	Isner	16.67	Kamke	12.79
157–159	Nieminen	8.51	Dolgopolov	30.80	Nadal	9.27
160–162	Janowicz	17.46	Granollers	13.16	Thiem	20.00
163–165	Gasquet	12.22	Bolelli	27.22	Monfils	13.53
166–168	Berlocq	9.63	Ferrer	12.22	Struff	21.65
169–171	Garcia-Lopez	10.70	Matosevic	20.17	Cuevas	8.76
172–174	Mannarino	16.05	Murray	11.38	Verdasco	13.91
175–177	Young	15.36	Seppi	16.78	Bellucci	21.90
178–180	Lopez	23.04	Monaco	17.20	Fogonini	18.57
181–183	Kohschrieber	9.15	Michon	11.04	Sock	8.97

Data	Player	UEpH	Player	UEpH	Player	UEpH
184–186	Istomin	15.25	Anderson	11.04	Johnson	16.26
187–189	Lajovic	11.37	Karlovic	7.29	Mayer	14.29
190–192	Zopp	16.42	Haider-Maurer	8.97	Gabashvili	10.00
193–195	Tursunov	14.76	Raonic	22.65	Cilic	20.10
196–198	Federer	12.51	Simon	7.96	Djokovic	11.62
199–201	Tsonga	8.93	Stepanek	14.62	Robredo	3.33
202–204	Janowicz	18.84	Gulbis	15.13	Isner	22.42
205–207	Berdych	19.88	Nadal	4.38	Monfils	16.47
208–210	Bautista Agut	13.25	Mayer	15.77	Fognini	23.82
211–213	Seppi	18.17	Garcia-Lopez	12.95	Karlovic	9.77
214–216	Ferrer	16.90	Young	15.07	Anderson	0.00
217–219	Sock	15.67	Granollers	5.67	Gasquet	12.71
220–222	Lajovic	8.96	Klizan	21.26	Verdasco	10.68
223–225	Kohlschrieber	13.36	Tsonga	25.62	Gulbis	14.32
226–228	Murray	13.36	Djokovic	12.13	Federer	15.95
229–231	Berdych	7.03	Raonic	15.79	Nadal	12.26
232–234	Isner	15.68	Granollers	3.68	Lajovic	27.74
235–237	Garcia-Lopez	24.41	Anderson	14.04	Verdasco	15.52
238–240	Monfils	14.24	Ferrer	8.42	Murray	17.24
241–243	Raonic	16.90	Berdych	12.61	Monfils	18.77
244–246	Djokovic	8.03	Gulbis	8.57	Murray	14.77
247–249	Nadal	12.08	Gulbis	17.14	Nadal	9.00
250–252	Ferrer	19.48	Djokovic	9.74	Murray	15.60
253	Nadal	10.81				
254	Djokovic	13.93				

Table A.12 4×200M freestyle relay times in the 2012 Olympics. Data for activity in Section 7.3.4

Nation	Men	Women
AUS	7:10.50	7:44.41
AUT	7:17.94	–
BEL	7:14.44	–
CAN	7:15.22	7:50.65
CHN	7:06.30	7:53.11
DEN	7:15.04	–
ESP	–	7:54.59
FRA	7:02.77	7:47.49
GBR	7:09.33	7:52.37
GER	7:06.59	7:58.93
HUN	7:11.64	7:54.58
ITA	7:12.69	7:52.75
JPN	7:11.74	7:54.56
NZL	7:17.18	7:55.92
POL	–	8:13.76
RSA	7:09.65	–
RUS	7:11.86	7:56.50
SLO	–	8:04.69
UKR	–	8:12.67
USA	6:59.70	7:42.92

International Olympic Committee. "Swimming." https://www.olympic.org/london-2012/swimming (accessed: December 3, 2018).

Table A.13 2016 Kentucky Derby horses. Data for activity in Section 8.3.4

Horse	Jockey	Trainer	Color	Birthplace	Sex
Nyquist	Mario Gutierrez	Doug F. O'Neil	Bay	Kentucky	Colt
Exaggerator	Kent J. Desormeaux	J. Keith Desormeaux	Dark bay	Kentucky	Colt
Gun Runner	Florent Geroux	Steven M. Asmussen	Chestnut	Kentucky	Colt
Mohaymen	Junior Alvarado	Kiaran P. McLaughlin	Gray	Kentucky	Colt
Suddenbreakingnews	Luis S. Quinonez	Donne K. Von Hemel	Bay	Kentucky	Gelding
Destin	Javier Castellano	Todd A. Pletcher	Gray	Kentucky	Colt
Brody's Cause	Luis Saez	Dale L. Romans	Bay	Kentucky	Colt
Mo Tom	Corey J. Lanerie	Thomas M. Amoss	Dark bay	Kentucky	Colt
Lani	Yutaka Take	Mikio Matsunaga	Gray	Kentucky	Colt
Mor Spirit	Gary L. Stevens	Bob Baffert	Dark bay	Pennsylvania	Ridgeling
My Man Sam	Irad Ortiz, Jr.	Chad C. Brown	Bay	Kentucky	Colt
Tom's Ready	Brian Joseph Hernandez, Jr.	Dallas Stewart	Dark bay	Pennsylvania	Colt
Creator	Ricardo Santana, Jr.	Steven M. Asmussen	Gray	Kentucky	Colt
Outwork	John R. Velazques	Todd A. Pletcher	Bay	Kentucky	Colt
Danzing Candy	Mike E. Smith	Clifford W. Sise, Jr.	Dark bay	Kentucky	Colt
Trojan Nation	Aaron T. Gryder	Patrick Gallagher	Bay	Kentucky	Colt
Oscar Nominated	Julien R. Leparoux	Michael J. Maker	Chestnut	Kentucky	Colt
Majesto	Emisael Jaramillo	Gustavo Delgado	Bay	Kentucky	Ridgeling
Whitmore	Victor Espinoza	Ron Moquett	Chestnut	Kentucky	Gelding
Shagaf	Joel Rosario	Chad C. Brown	Bay	Kentucky	Colt

Kentucky Derby. "Horses." http://www.kentuckyderby.com/horses (accessed: August 10, 2016).

Glossary

Baseball Today's game of baseball has its roots in the game of rounders (a mix of baseball with dodgeball) played in early England. It became an organized sport in the mid 19th century in America. The Cincinnati Reds were the first professional team in any sport in the United States, and now Major League Baseball consists of 30 teams in the United States and Canada. Although declining in popularity in recent times, baseball still has worldwide support, especially in North America, the Caribbean, and Japan, and continues to be played from primary school Little Leagues up through college teams and the pros.[1,2]

Basketball Basketball was invented by James Naismith in 1891 at what is now Springfield College in Massachusetts. It became popular at colleges across the country and spread across the world, with the first amateur international organization starting in 1932. The first professional organization, the National Basketball Association (NBA), was established in 1946 in the United States, and the Women's National Basketball Association (WNBA) began in 1996. Internationally, more than 200 countries have men's and women's national teams, which compete in various events including the Olympics and the Basketball World Cup.[3,4,5,6]

Cycling The bicycle was invented in the early 1800s and was used for recreation and transportation before bicycle racing became an organized sport in the mid 1800s. The governing body for cycling, the Union Cycliste Internationale (UCI), was established in 1900 and recognizes several disciplines, including road and track races, as well as mountain and BMX events. The UCI organizes many popular road races including the Tour de France, a team event encompassing three weeks in July which attracts millions of viewers each year.[7,8]

[1] America's Story. "Play Ball!" http://www.americaslibrary.gov/jp/bball/jp_bball_subj.html (accessed: December 5, 2018).

[2] NPR. "The 'Secret History' of Baseball's Earliest Days." Fresh Air. https://www.npr.org/2011/03/16/134570236/the-secret-history-of-baseballs-earliest-days (accessed: December 5, 2018).

[3] Springfield College. "Where Basketball was Invented: The History of Basketball." https://springfield.edu/where-basketball-was-invented-the-birthplace-of-basketball (accessed: December 5, 2018).

[4] International Basketball Federation. "FIBA.basketball." http://www.fiba.basketball/ (accessed: December 5, 2018).

[5] Sky Sports, "NBA." http://www.nba.com/ (accessed: December 5, 2018).

[6] Women's National Basketball Association. https://www.wnba.com/ (accessed: December 5, 2018).

[7] Union Cycliste Internationale. http://www.uci.ch/ (accessed: December 5, 2018).

[8] National Museum of American History. "America on the Move." Smithsonian. http://amhistory.si.edu/onthemove/themes/story_69_2.html (accessed: December 5, 2018).

Football Traditionally, football has been a uniquely American game with little worldwide interest. With roots in the games of soccer and rugby, football was loosely organized in the 19th century, and the first college football game was played between Rutgers and Princeton in 1869. Professional football games soon followed, with the first professional league established in 1920. The modern-day National Football League was formed in 1966, the year of the first Super Bowl. With this event, college football bowl games, and regular-season games, football is easily the most watched sport in the United States, with teams known for having large, enthusiastic fanbases.[9,10,11,12]

Golf Originating in Scotland in the 15th century, golf is one of the older games played professionally. In 1754, the first society was formed at Saint Andrews, and the first Open Championship was played in 1860 in Scotland. Men's and women's US Amateur Championships were played for the first time in 1895, and in 1901 the Professional Golfers' Association (PGA), the current governing body, was established. Today golfers in the PGA compete yearly in four major tournaments, namely, the Masters tournament, the US Open, the Open Championship, and the PGA Championship, while golfers from the LPGA, the Ladies Professional Golf Association, have five masters tournaments, namely, the ANA Inspiration, the US Women's Open, the Women's PGA Championship, the Women's British Open, and the Evian Championship.[13,14,15]

Horse racing Thoroughbred horse racing began in England in the 1700s after Arabian horses were imported from the Middle East. It is popular in many parts of the world today, with casual fans to serious gamblers hoping to win wagers. Three of the most important thoroughbred races being run each year, known as the Triple Crown, are the Kentucky Derby, the Preakness Stakes, and the Belmont Stakes. The National Thoroughbred Racing Association (NTRA) acts as the governing body for the sport.[16,17]

Olympic Games The modern Olympic Games have been held since 1896 and involve a variety of sports and disciplines. Countries are invited to send their national athletes to compete for gold, silver, and bronze medals. Since 1992 the Olympics have been held every two years, alternating between the Winter Olympics, with ice and snow events, and the larger Summer Olympics, encompassing track and field, swimming, gymnastics, basketball, and soccer competitions, among others.[18]

[9] Pro Football Hall of Fame. "Football History." www.profootballhof.com/football-history/history-of-football/ (accessed: December 5, 2018).

[10] Norman, Jim. "Football Still Americans' Favorite Sport to Watch." Gallup. January 4, 2018. http://news.gallup.com/poll/224864/football-americans-favorite-sport-watch.aspx (accessed: December 5, 2018).

[11] National Football League. https://www.nfl.com/ (accessed: December 5, 2018).

[12] NCAA. "Football." https://www.ncaa.com/sports/football/fbs (accessed: December 5, 2018).

[13] International Golf Federation. "History of Golf." https://www.igfgolf.org/about-golf/history/ (accessed: December 5, 2018).

[14] Ladies Professional Golf Association. http://www.lpga.com/ (accessed: December 5, 2018.)

[15] PGA Tour. https://www.pgatour.com/ (accessed: December 5, 2018).

[16] American Museum of Natural History. "Sport of Kings." https://www.amnh.org/exhibitions/horse/how-we-shaped-horses-how-horses-shaped-us/sport/sport-of-kings/ (accessed: December 5, 2018).

[17] National Thoroughbred Racing Association. https://www.ntra.com/ (accessed: December 5, 2018).

[18] International Olympic Committee. "Athens 1896." https://www.olympic.org/athens-1896 (accessed: December 5, 2018).

Skiing Skiing is a recreational sport that is also well known competitively by its inclusion in the winter Olympic Games. Skiing largely originated in Asia and Scandinavia as a method of transportation in prehistoric times. Today millions of skiers and snowboarders visit resorts across the world each year. The Fédération Internationale de Ski (FIS) recognizes several disciplines in competitions, including Alpine, Nordic, and freestyle skiing, and holds frequent events throughout the year, including World Cups in these disciplines and other snow-related sports.[19,20]

Soccer Easily the most popular sport worldwide, association football (soccer) began in England in 1863 after splitting from the proponents of rugby-style football. Organized competition followed, with the FA (Football Association) Cup being held in 1872, and the sport was professionalized in the late 19th century. The Fédération Internationale de Football Association (FIFA) the world governing body, was established in 1904. Today FIFA consists of over 200 men's and women's national association teams and sponsors the world's most watched sporting event, the FIFA World Cup, a competition between 32 qualifying teams held every four years.[21,22]

Swimming When thinking of swimming as a sport, many people will first recall the events in the Olympics, but swimming has a strong presence outside of the Olympics in youth and college competitions as well as professional events. A recreational sport for many, the first competitions were held in the 1880s, and swimming was a part of the first modern Olympic Games in 1896. The sport's governing body, the Fédération Internationale de Natation (FINA), currently recognizes four different strokes: backstroke, breaststroke, butterfly, and freestyle. In competitions, men and women swim these strokes individually or in relays of different distances.[23,24]

Tennis Tennis evolved from an 11th century French sport known as "jeu de palme," and the more modern game was established in England at the end of the 19th century. The first men's competition at Wimbledon was held in 1877, while the first women's match championship at Wimbledon occurred in 1884. Tennis major championships may be played on grass, clay, or hard courts and its four major events, known as the Grand Slam, are the Australian Open, the French Open, Wimbledon, and the US Open. The sport became professional in 1968, with the Association of Tennis Professionals (ATP) and the Women's Tennis Association (WTA) as the sponsoring organizations.[25,26,27]

Track and Field The origins of organized track and field events can be traced back to the first Olympic Games in ancient Greece. It it very natural for athletes to wish to determine who can run the fastest, jump the highest, or throw a spear the farthest. Modern track and field includes

[19] Fédération Internationale de Ski. http://www.fis-ski.com/ (accessed: December 5, 2018).

[20] Fessl, Sophie. "A Brief History of Skis." JSTOR Daily. https://daily.jstor.org/brief-history-skis/ (accessed: December 5, 2018).

[21] Worldatlas. "The Most Popular Sports in the World." https://www.worldatlas.com/articles/what-are-the-most-popular-sports-in-the-world.html (accessed: December 5, 2018).

[22] FIFA. "History of Football—The Origins. FIFA, Fédération Internationale de Football Association. http://www.fifa.com/about-fifa/who-we-are/the-game/index.html (accessed: December 5, 2018).

[23] Fédération Internationale de Natation. http://www.fina.org/ (accessed: December 5, 2018).

[24] International Olympic Committee. "Swimming." https://www.olympic.org/swimming (accessed: December 5, 2018).

[25] International Olympic Committee. "A Brief History of Tennis." https://www.olympic.org/news/a-brief-history-of-tennis (accessed: December 5, 2018).

[26] ATP Tour. http://www.atpworldtour.com/ (accessed: December 5, 2018).

[27] WTA. http://www.wtatennis.com/ (accessed: December 5, 2018).

versions of these events, with short- and long-distance foot races, high jump and long jump, javelin, and shot put, along with other disciplines and combined events. Athletes may compete in team and individual events. Today men's and women's major track and field competitions are held biannually, organized by the International Association of Athletics Federations (IAAF).[28]

---

[28] International Association of Athletics Federations. "IAAF World Championships." https://www.iaaf.org/competitions/iaaf-world-championships (accessed: December 5, 2018).

Author Biographies

Tricia Muldoon Brown is an Associate Professor at Georgia Southern University. She has a B.A. in mathematics from Marshall University and an M.S. and Ph.D. in mathematics from the University of Kentucky. Her research interests include studying the connections between commutative algebra, combinatorics, simplicial topology, enumerative combinatorics, and recreational mathematics. With her collaborators, she has published pure mathematics articles in journals such as *Discrete Mathematics*, the *Journal of Combinatorics*, and the *Journal of Algebra and Its Applications*, book chapters in the Mathematical Association of America (MAA) text *Mathematics and Sports* and in the *Handbook of Research on Active Learning and the Flipped Classroom Model in the Digital Age*, and a forthcoming chapter in *Mathematics and Social Justice: Perspectives and Resources for the College Classroom*. She is actively involved in her teaching and was a 2009 Project NExT fellow and a 2015 Georgia Governor's Teaching Fellow, and was awarded a 2016 Affordable Learning Georgia Textbook Transformation Grant to adapt an open-source introductory mathematics textbook for use on her campus.

Eric B. Kahn joined the faculty at Bloomsburg University in 2009 as an Assistant Professor of Mathematics and earned tenure and promotion to Associate Professor in the spring of 2014. He received his B.A. in mathematics from Kenyon College in 2004 and his M.A. and Ph.D. degrees in mathematics from the University of Kentucky in 2006 and 2009, respectively. He has published research articles in the peer-reviewed journals *Rocky Mountain Journal of Mathematics* and *Communications in Algebra*, as well as more pedagogically minded articles in *PRIMUS* and the MAA's book *Mathematics and Sports*. In 2015, his textbook *Problems and Proofs in Numbers and Algebra*, which he coauthored with Dr. Richard Millman, then of the Georgia Institute of Technology, and Peter Shuie, of the University of Nevada, Las Vegas, was published by Springer. He is an active member of the MAA, was a Project NExT fellow in the 2009 cohort, was elected to serve a two-year term starting in 2017 as Program Coordinator for the SIGMAA on Inquiry Based Learning, and, since 2010, has co-coordinated or coordinated the student paper competition of his local MAA section, EPaDel.

Index by Mathematics

Index by Sport

Sport	Activities	Projects
Baseball	3.3.1, 6.3.3, 7.3.1	3.4.1, 6.4.2, 7.4.2
Basketball	3.3.2, 4.3.3, 5.3.3, 5.3.4, 5.3.5	3.4.3, 5.4.1
Cycling	2.3.1, 2.3.2	2.4.1
Fans	5.3.2, 8.3.1, 8.3.5	8.4.2
Football	3.3.4, 3.3.5, 4.3.4, 4.3.5, 8.3.3, 8.3.5	4.4.3, 8.4.3
Golf	6.3.4, 6.3.5	6.4.3
Horse Racing	8.3.4	8.4.1
Olympics	2.3.5, 3.3.3, 5.3.1, 5.3.2, 6.3.2, 7.3.4	2.4.3, 3.4.2, 5.4.2, 5.4.3, 6.4.1, 7.4.3
Skiing	2.3.3, 2.3.4	2.4.2
Soccer	4.3.1, 4.3.2, 8.3.2	4.4.1, 4.4.2
Swimming	7.3.4, 7.3.5	7.4.3
Tennis	7.3.2, 7.3.3	7.4.1
Track and Field	6.3.1, 6.3.2	6.4.1